AF737053

Predictive Modelling for Football Analytics

Predictive Modelling for Football Analytics discusses the most well-known models and the main computational tools for the football analytics domain. It further introduces the footBayes R package that accompanies the reader through all the examples proposed in the book. It aims to be both a practical guide and a theoretical foundation for students, data scientists, sports analysts, and football professionals who wish to understand and apply predictive modelling in a football context.

Key Features

- Discusses various modelling strategies and predictive tools related to football analytics
- Introduces algorithms and computational tools to check the models, make predictions, and visualize the final results
- Showcases some guided examples through the use of the footBayes R package available on CRAN
- Walks the reader through the full pipeline: from data collection and preprocessing, through exploratory analysis and feature engineering, to advanced modelling techniques and evaluation
- Bridges the gap between raw football data and actionable insights

This text is primarily for senior undergraduates, graduate students, and academic researchers in the fields of mathematics, statistics, and computer science willing to learn about the football analytics domain. Although technical in nature, the book is designed to be accessible to readers with a background in statistics, programming, or a strong interest in sports analytics. It is well-suited for use in academic courses on sports analytics, data science projects, or professional development within football clubs, agencies, and media organizations.

CHAPMAN & HALL/CRC DATA SCIENCE SERIES

Reflecting the interdisciplinary nature of the field, this book series brings together researchers, practitioners, and instructors from statistics, computer science, machine learning, and analytics. The series will publish cutting-edge research, industry applications, and textbooks in data science.

The inclusion of concrete examples, applications, and methods is highly encouraged. The scope of the series includes titles in the areas of machine learning, pattern recognition, predictive analytics, business analytics, Big Data, visualization, programming, software, learning analytics, data wrangling, interactive graphics, and reproducible research.

Recently Published Titles

Soccer Analytics
An Introduction Using R
Clive Beggs

Spatial Statistics for Data Science
Theory and Practice with R
Paula Moraga

Research Software Engineering
A Guide to the Open Source Ecosystem
Matthias Bannert

The Data Preparation Journey
Finding Your Way With R
Martin Hugh Monkman

Getting (more out of) Graphics
Practice and Principles of Data Visualisation
Antony Unwin

Introduction to Data Science
Data Wrangling and Visualization with R Second Edition
Rafael A. Irizarry

Data Science
A First Introduction with Python
Tiffany Timbers, Trevor Campbell, Melissa Lee, Joel Ostblom and Lindsey Heagy

Mathematical Engineering of Deep Learning
Benoit Liquet, Sarat Moka, and Yoni Nazarathy

Introduction to Classifier Performance Analysis with R
Sutaip L.C. Saw

Predictive Modelling for Football Analytics
Leonardo Egidi, Dimitris Karlis, and Ioannis Ntzoufras

For more information about this series, please visit: https://www.routledge.com/Chapman--HallCRC-Data-Science-Series/book-series/CHDSS

Predictive Modelling for Football Analytics

Leonardo Egidi, Dimitris Karlis, and
Ioannis Ntzoufras

CRC Press
Taylor & Francis Group
Boca Raton London New York

CRC Press is an imprint of the
Taylor & Francis Group, an **informa** business
A CHAPMAN & HALL BOOK

First edition published 2026
by CRC Press
2385 NW Executive Center Drive, Suite 320, Boca Raton FL 33431

and by CRC Press
4 Park Square, Milton Park, Abingdon, Oxon, OX14 4RN

CRC Press is an imprint of Taylor & Francis Group, LLC

© 2026 Leonardo Egidi, Dimitris Karlis, and Ioannis Ntzoufras

Reasonable efforts have been made to publish reliable data and information, but the author and publisher cannot assume responsibility for the validity of all materials or the consequences of their use. The authors and publishers have attempted to trace the copyright holders of all material reproduced in this publication and apologize to copyright holders if permission to publish in this form has not been obtained. If any copyright material has not been acknowledged, please write and let us know so we may rectify in any future reprint.

Except as permitted under U.S. Copyright Law, no part of this book may be reprinted, reproduced, transmitted, or utilized in any form by any electronic, mechanical, or other means, now known or hereafter invented, including photocopying, microfilming, and recording, or in any information storage or retrieval system, without written permission from the publishers.

For permission to photocopy or use material electronically from this work, access www.copyright.com or contact the Copyright Clearance Center, Inc. (CCC), 222 Rosewood Drive, Danvers, MA 01923, 978-750-8400. For works that are not available on CCC, please contact mpkbookspermissions@tandf.co.uk

Trademark notice: Product or corporate names may be trademarks or registered trademarks and are used only for identification and explanation without intent to infringe.

ISBN: 978-1-032-03064-7 (hbk)
ISBN: 978-1-032-03063-0 (pbk)
ISBN: 978-1-003-18649-6 (ebk)

DOI: 10.1201/9781003186496

Typeset in CMR10 font
by KnowledgeWorks Global Ltd.

Publisher's note: This book has been prepared from camera-ready copy provided by the authors.

To our families
and the football fans in love with statistics.

Contents

Preface

In recent years, football—known globally as the beautiful game—has undergone a profound transformation shaped by the rise of data and analytics. What was once the exclusive domain of scouts' intuition and coaches' experience has been augmented, and in many ways revolutionized, by the advent of predictive modelling and data-driven decision-making. *Predictive Modelling for Football Analytics* emerges at this intersection, where the love for the game meets the rigour of statistical science.

The impetus for this book stems from a growing demand to bridge the gap between raw football data and actionable insights. While statistics have long been used in other sports such as baseball and basketball, the inherently fluid and low-scoring nature of football has posed unique challenges for analysts and modellers. Our motivation in writing this book was to provide a structured approach to tackling these challenges using a comprehensive suite of predictive tools and methodologies—grounded in machine learning, statistical inference, and domain-specific knowledge.

This book aims to be both a practical guide and a theoretical foundation for students, data scientists, sports analysts, and football professionals who wish to understand and apply predictive modelling in a football context. It walks the reader through the full pipeline: from data collection and preprocessing, through exploratory analysis and feature engineering, to advanced modelling techniques and evaluation. Emphasis is placed on reproducibility, interpretability, and real-world application, with each chapter grounded in examples and datasets reflective of contemporary football.

Although technical in nature, the book is designed to be accessible to readers with a background in statistics, programming, or a strong interest in sports analytics. It is well-suited for use in academic courses on sports analytics, data science projects, or professional development within football clubs, agencies, and media organizations.

A work of this nature is never a solitary effort. We would like to express our gratitude to the many individuals and institutions that supported us throughout this journey. Special thanks go to our colleague, professor Nicola Torelli, and to some researchers and practitioners in the football analytics community, Roberto Macrì Demartino and Vasilis Palaskas, whose insights, footBayes R package development, and critical discussions have shaped much of the field—and this book.

We also extend our appreciation to friends and mentors who provided feedback during the writing process, and to our families for their patience and encouragement.

It is our hope that this book not only informs but inspires further exploration and innovation in football analytics. Whether you are building models for match outcomes, scouting promising talent, or optimizing team performance, may this work serve as a valuable resource in your pursuit of insight.

Leonardo Egidi, Ioannis Ntzoufras, and Dimitris Karlis

About the Authors

Leonardo Egidi is a distinguished statistician and academic, recognized for his significant contributions to the fields of Bayesian statistics, sports analytics, and statistical modeling. He is an associate professor of statistics at the University of Trieste, where his research primarily focuses on applying advanced statistical methods to real-world problems, with a particular emphasis on sports data analysis, genomics, and predictive modeling.

Professor Egidi is well-known for his work in theoretical Bayesian inference and in football analytics, particularly in the development of models to predict match outcomes, assess player performance, and optimize team strategies. His research includes the application of machine learning algorithms and Bayesian methods to enhance the accuracy of predictions and provide insights into various aspects of the game. He has published extensively in leading academic journals and has collaborated with both academic researchers and sports organizations to advance the field of sports data science.

In addition to his work on football, Professor Egidi has also contributed to statistical methodology in other domains, including economics, biostatistics, and social sciences. His expertise lies in the integration of complex data structures, such as hierarchical models, into practical solutions that can drive decision-making processes.

Beyond his research, Leonardo Egidi is actively involved in teaching and mentoring, fostering the next generation of statisticians and data scientists. His work has made a substantial impact on both the academic community and the sports industry, cementing his reputation as a leading figure in the application of statistics to sports analytics.

He is an associate editor for the *Journal of Quantitative Analysis in Sports* and the creator and the maintainer of the CRAN R package *footBayes.*

Dimitris Karlis is a distinguished statistician and academic widely recognized for his contributions to the fields of statistical modeling, discrete valued time series analysis, model-based clustering and sports analytics. He is a full professor at the Athens University of Economics and Business, where his research focuses on the development and application of advanced statistical methods for various problems and disciplines. He has served as director of the

MSc in Statistics program at AUEB, (2019-today), Director of the Laboratory of Computational and Bayesian Statistics (2017–today) and vice-president of the Research Committee of AUEB (2019 -today). Professor Karlis has made significant contributions to the statistical analysis of sports data, especially in football (soccer), basketball, handball, and other team sports. His work on modeling match outcomes, player performance, and team strategies has had a substantial impact on both academic research and practical applications in the sports industry. He is known for pioneering methods such as the use of generalized linear models and mixed-effects models for analyzing sports data as well as the development of innovative models for various sports.

Professor Karlis has served as Editor-in-chief for *Journal of Quantitative Analysis in Sports.* He has been an Associate Editor in other journals including *Stochastic Environmental Research* and *Risk Assessment*, *Metron,* and *Risks.* He has been a Guest Editor in a special issue on Mathematics in Sports in *IMA Mathematics for Management Journal.*

Throughout his career, Professor Karlis has authored more than 120 peer-reviewed journal articles in leading statistical journals, and he has collaborated with various sports organizations, data analysts, and bookmakers to apply statistical methods to real-world problems. His research has improved predictive modeling in sports, offering valuable insights into game outcomes, player behavior, and tactical decisions.

Beyond his research, he is actively involved in teaching and mentoring.

He has authored two textbooks in Statistics. In addition to his academic work, Dimitris Karlis is an active advisor for the use of data science and statistics in decision-making processes across different industries, including sports, finance, real estate, and healthcare. His academic and professional achievements have made him a key figure in the development of statistical techniques and their applications in both research and industry.

Ioannis Ntzoufras is a distinguished statistician and academic, widely recognized for his contributions to statistical modeling, Bayesian analysis, and sports analytics. He is a full professor in the Department of Statistics at the Athens University of Economics and Business (AUEB). He is particularly known for his work in Bayesian statistics, including the development and application of Markov Chain Monte Carlo (MCMC) methods and Bayesian variable selection techniques. His research also addresses computational strategies and prior formulation for Objective Bayesian model comparison. These methodologies have been applied across various domains, with a strong emphasis on sports analytics—especially in football (soccer).

He served as Head of the Department of Statistics at AUEB from 2020 to 2025. He was awarded the **Lefkopoulion Award** by the Greek Statistical Institute in 2000 and is the author of the acclaimed book *Bayesian Modeling Using WinBUGS* (Wiley), which received an honorable mention in Mathematics at the 2009 PROSE Awards. In addition, he has authored a Greek-language textbook titled *Introduction to Programming and Statistical Data Analysis with R*, and he has served as the scientific editor for the Greek translations of two influential texts: Andy Field's *Discovering Statistics with R* and Bernard Rosner's *Fundamentals of Biostatistics.*

Professor Ntzoufras has served as an associate editor for several journals, including the *Journal of the Royal Statistical Society C*, *Statistics*, and the *Journal of Quantitative Analysis in Sports.* As of April 2025, Professor Ntzoufras has authored 76 peer-reviewed journal articles, accumulating over 6100 citations and an h-index of 29 on Google Scholar. He remains actively engaged in research, with current projects focusing on Bayesian methodology, variable selection, applied statistics, biostatistics, psychometrics, and sports analytics. His contributions to sports analytics have led to the creation of models that enhance performance prediction and strategic planning in football, basketball, and volleyball.

1

A short introduction to football analytics

1.1 Introduction

Football is the most popular sport in the world, with an estimated amount of 3.5 billion fans and 250 million players worldwide; see for example in Szymanski (2003) and Giulianotti and Robertson (2004). The sport is played in almost every part of the world, and major football events such as the World Cup attract enormous numbers of spectators and media attention possibly more than any other sport event in the world. The football-related industry has a significant impact, and its economic role is also relevant; among the others, betting results in football matches are also very popular, while the value of the soccer betting market is growing.

This growing impact on the economy and also on the daily life of the world has also increased interest for scientific and academic study of football. Apart from being a popular entertainment, football is still a sport which also attracts a large global athlete community. A thriving scientific field has resulted from the necessity for academic study, research, and comprehension of sports, and football in particular.

One of the most important aspects about football is its inherent uncertainty leading to unpredictability (Scarf et al., 2022), which is the result of the fact that football is, in fact, a low-score sport. Typically we expect, depending on the tournament or league, around two or three goals per match, while other popular sports like basketball have scores up to 200 per match in total. The small expected number of goals per match and the high uncertainty of the final football outcome lead to more unexpected or surprising results, and, therefore, increase the excitement and thrill among fans. For this purpose, there have been many efforts to comprehend and model the result of a football match along with the interest of modelling other, specific, aspects of the game. Thus, the literature on football modelling is growing rapidly. Currently, the application of machine learning methods and statistical models for predicting football match results is very popular among researchers, in both academic and industrial settings, not only for the possibility of financial gains but also for the difficulties that such modelling or prediction exercise poses.

Football modelling involves various stakeholders. To start with, the athletic teams are interested in evaluating their performances, detecting their weaknesses and identifying the factors that contribute to their success based

DOI: 10.1201/9781003186496-1

on data. Football tactics are also crucial, as well as using data to assess the performance of players or to devise better strategies during the game. For instance, there is considerable debate about the optimal substitution strategy (Silva and Swartz, 2016; Myers, 2012). Furthermore, injury prevention is crucial in professional soccer (and sports, more generally) due to the high cost of recovery for players and the significant impact of injuries on a club's performance (Rossi et al., 2017). Not all trainers or teams are willing to invest in data analytic methods, but it is evident that as time passes, more and more teams are adopting such a data-based holistic approach to generate insights that may help improve their performance

Another target group for the implementation and use of athletic data analytics is represented by the football spectators and fans. They are interested in understanding the game, predicting the outcome of their team and also engaging in betting activities for fun or for profit. Therefore, they require to have access to a deeper analysis of the sport based on the generated data and facts. Often, they also enjoy following and evaluating the performance and the career of specific (usually their favourite) players, as well as to get involved in fantasy football games. Fantasy football is a popular game in which the fan can act as the owner of a team. Under this role, he is invited to select his own roster from a list of actual players in the league of interest under a specific budget. After the selection of his team roster, he competes against other fantasy players by collecting score points based on the performance of the chosen players in real football games. This can also involve betting in some countries. As you may understand, fantasy football has its own modelling techniques (Egidi and Gabry, 2018) with the analysis focused on the collected points by each player or on smaller events (such as yellow or red cards) that may attribute or remove points from the score of each player.

A third relevant stakeholder is represented by the betting companies. They require models to set their betting odds, although the betting market itself will influence their decisions later on. Typically, one game can generate a large number of different and sometimes unrelated bets, which demand very specialized modelling approaches, different in nature, from predicting the score or only the outcome of the score difference.

Another group consists of journalists—including TV broadcasters—who need to examine deeper the data to comprehend the performances of teams or individual players, visualize them in a clear and appealing way, and create a more holistic view of the teams. Moreover, the academia, perhaps wearing many of the previous hats, is especially interested in developing new models and creating new insights based on football data. For the end, we have left the sports-related market that needs to gain insights from the game in order to create new products and services. All the above constitute the general ecosystem of users of football models and also highlight the reasons for their growing popularity in the academic agenda. Over the last 50 years, there has been a considerable qualitative improvement regarding the football data availability. Scores of football matches have been widely available, in several sources, shortly after the matches. However, since then, the available data have become more detailed and comprehensive. A rough timeline of the data availability development is as follows:

- The first attempt to collect specific football data was by Reep and Benjamin (1968) who used manual and personal observation methods. However, such data are likely to be incomplete and inaccurate. Moreover, by this way, it is difficult to cover a large number of matches or a large variety of events.

- Later on, the wide broadcasting of football has led to a growing interest of the analysis of the data related to this sport. Hence, more detailed data were collected and were made available after the end of each match in the form of box-scores. A typical box-score can include various measures that capture specific playing characteristics of each team (or player) such as minutes played, ball possession time, cards received, fouls and corners committed, among others.

- The next step in the development of recorded football data was the availability of specific, specialized, in-play event data including their timestamp (i.e. the exact time that each event appeared in the match); see for example in Anzer et al. (2021). Event data and positional data from football matches aim to record all events and movements on the football pitch and are extensively studied in sports science. Event data provide a detailed and ordered sequence of all the player's actions during the match, such as passes, shots, or tackles. Each event is characterized by the time and location where the action occurred on the field as well as the event type. Depending on the data provider, additional information such as a subtype or the outcome of the event may be given. Data are mostly collected manually by video analysis, but, lately, there are attempts to automatically detect them through artificial intelligence (AI) visualization tools.

- Recently, tracking data are available from devices that can record the exact location of the players and the ball almost 20 times per second. Such data can reveal detailed information about the players, the distance run by them, their performance and the tactical elements of players and teams, providing an extremely detailed picture of the team and its fitness condition at each point of the game. Such data, although not yet widely available, can transform the way we view the sport, and they can offer new exciting opportunities for implementing sophisticated statistical models algorithms (Goes, 2021).

This kind of data is often accompanied with data related to the odds offered by betting companies, which can largely be interpreted as implied probabilities for the outcome. Such information can be used, mainly in Bayesian learning methods, as a source of information from experts about the match which can increase the accuracy of the implemented methods and which can increase the accuracy of the implemented methods and improve the matches' understanding. algorithms

It is evident that football analytics research is growing rapidly. In the following section, we will attempt to provide some general directions of the relevant cutting-edge research in the field.

1.2 The early years of statistical analysis of football data

Research on football has a much shorter history than other sports like baseball and basketball. An early attempt was made by Moroney (1956), who used the Poisson distribution to fit the number of goals scored in a football match. This was rather an illustration about the use of Poisson distribution on football data rather than a deep data-based study of the game. The underlying assumption was that if pure chance governed the outcome of a football game, the Poisson distribution would be able to fit the number of goals. This assumption seems rather simplistic and unrealistic for football, as it implies that all teams in a league have the same strength/ability and also that there is no correlation between the two teams competing during the game, which is counter-intuitive since the two teams interact. However, the use of the Poisson distribution has been a very important first step for football research, since, nowadays, it is the initial starting point from which we can build more sophisticated models.

Reep and Benjamin (1968) is perhaps the first paper that attempts to provide insights into football by using models beyond the Poisson for certain aspects of the game. The paper studied the number of successful passes for one team, which are followed by (a) a shot either a shot at the goalpost, (b) an infringement, or (c) an intercepted pass attempt. The negative binomial distribution was used as an indication that other factors besides chance affect the mechanism, such as individual abilities. The negative binomial was also implemented to the number of goals in Reep et al. (1971). Nowadays, this distribution complements the Poisson distribution as one of the initial points of statistical research in football. These papers are the two pioneering research works on the notational skill in football, which also initiated discussion about direct game versus positional game (i.e. trying to attack as fast as possible versus waiting for an opportunity to arise in order to attack).

1.3 Modelling approaches

Modelling approaches for football were developed early in the research agenda. There are different approaches for this task, depending on the level of information that one aims to model or predict. Do we only focus on modelling or predicting the outcome of the game (win/draw/loss) or the score itself? Or do we wish to predict the half-time score only, or both the final and half-time scores simultaneously? Thus, before providing more details, we need to ask two specific questions that their answer will lead to different categories of models.

1. **Outcome or score:** The first question we need to answer is whether the match outcome (win/draw/loss) or the detailed score (i.e. number of goals) should be modelled as the main outcome or response variable. The latter carries richer information since the goal difference also captures the volume of the dominance of the winner while the sum of the goals may imply a poor or more exciting game with respect to the main events of the game, which are the goals scored. Moreover, the final match outcome can be derived directly by the score. Scarf and Rangel Jr (2017) introduced the terms *direct* and *indirect* models for the two alternative modelling approaches. In Egidi and Torelli (2021a), this distinction is referred to as *result-based* versus *goal-based models.* The debate about the most appropriate approach is still ongoing and depends largely on the objective of such models. Note that there are models that aim to predict the score difference (Karlis and Ntzoufras, 2009) or even the time between goals. Such models may fall in a grey area between the above-mentioned broad categories of direct/indirect or result/goal-based models (Dixon and Robinson, 1998).

2. **prediction versus Exploration:** A second way to distinguish the published research work on football analysis is based on whether the aim is to predict the outcome beforehand or to identify the important factors that influence the outcome after the game. Predictive models usually rely on information which is available before the game based on the abilities of the teams up to that point—or some proxy like a rating score—and possibly some game-specific information such as betting odds. For the latter case, we can fit explanatory models that use information from the match itself, such as team statistics from the match box-score or even some socio-economic factors. For instance, Schauberger et al. (2018a) used on-field statistics from each match as covariates, such as the distance covered by players, ball possession, tackling success rate, shots on goal, completion rate, fouls suffered, offsides, and so on. Such variables are only available after the match, so they cannot be used prediction but they may reveal the features of the game that make a team successful. On the other hand, predictive models should use similar variables but from information gathered from the previous games (usually 5–10 previous games). The structure of the models is more or less the same, but the scope is different and therefore different information in incorporated as inputs/covariates.

In this section, we briefly discuss some of the most important models, their rationale, and their historical development. A more detailed and more mathematical description of the models will be given in Chapters 4 and 5.

1.4 Modelling the outcome

The match/game outcome is defined as a variable with three possible values: win, draw, or loss for the home team.

One of the earliest attempts to predict the match outcome was made by Stefani (1980), who used a least squares approach. This model was a simple normal regression with the goal difference as the response variable and the team ranking difference as the explanatory variable. The model predicted a win/draw/loss by considering positive values greater than 0.5 as a home win, values between −0.5 and 0.5 as a draw, and negative values below −0.5 as an away win. Although they model ignored certain characteristics of football, it was quite sufficient and important progress at that time. The logic of the models can be connected with a multinomial probit regression assuming a normal latent variable as response.

Related types of models are the paired comparison models, such as the Bradley-Terry model (Bradley and Terry, 1952). The Bradley-Terry paired comparison model was originally developed to associate the subjective preference of a set of objects when compared in pairs by one or more judges. This model has been applied to studies of preference and choice behaviour, as well as to the ranking of competitors and the prediction outcomes in sports such as chess, tennis, and soccer. The original model does not account for ties (draws).

Given a pair of individuals i and j drawn from some population, the model estimates the probability that team i will win against team j using

$$P_{ij} = \Pr(i > j) = \frac{B_i}{B_i + B_j},$$

where $P_{ij} = \Pr(i > j)$ is the probability of i team winning j team, and B_i is a positive real-valued score assigned to team i to represent its latent ability. This can be expressed as

$$\Pr(i > j) = \frac{e^{\beta i}}{e^{\beta i} + e^{\beta_j}} \quad \text{or} \quad \log \frac{\Pr(i > j)}{\Pr(j > i)} = \beta_i - \beta_j. \tag{1.1}$$

To apply the model two real data, one needs to estimate the parameters β_i that represent the abilities of the teams. The β's need to be constrained for identifiability purposes. It model provides a rating of the teams through the different estimated β's.

This model is essentially just a logistic regression. model Another approach, mainly used in chess, is the Thurstone-Mosteller model (Henery, 1992), which replaces the assumption of an underlying logistic distribution with a Gaussian distribution.

The Bradley-Terry (BT) model is suitable for binary outcomes and hence sports that always have a winner. Nevertheless, it needs to be modified for football in order to account for the possibility of a draw. There are various ways to do this. One possible model that allows for draws assumes is the one proposed by Rao and Kupper (1970), which introduces an extra parameter $\theta > 0$ in the following way

$$\begin{aligned} P_{ij} = \Pr(i > j) &= \frac{B_i}{B_i + \theta B_j}, \\ P_{ji} = \Pr(j > i) &= \frac{Bj}{B_j + \theta B_i}, \\ \Pr(i = j) &= \frac{B_i B_j(\theta^2 - 1)}{(B_j + \theta B_i)(Bi + \theta Bj)}, \end{aligned}$$

or alternatively, the model ties introduced by Davidson (1970) given by

$$\begin{aligned} P_{ij} = \Pr(i > j) &= \frac{B_i}{B_i + Bj + \nu\sqrt{B_i B_j}}, \\ P_{ji} = \Pr(j > i) &= \frac{B_j}{Bi + B_j + \nu\sqrt{BiB_j}}, \\ \Pr(i = j) &= \frac{\nu\sqrt{B_i B_j}}{Bi + Bj + \nu\sqrt{B_i B_j}} \end{aligned}$$

for $\nu \geq 0$. These two models introduce an additional parameter (θ or ν) to reduce the probability of the two outcomes (win/loss) and attribute it to the third outcome of draw. Note that we can obtain the original Bradley-Terry model (without a draw) for $\theta = 1$ and $\nu = 0$.

Before these two extensions, Glenn and David (1960) modified the Thurstone-Mosteller model to allow small differences to become ties, while many years later, Kuk (1995) applied the approach model to football. For more extensions of the Bradley-Terry models, see Baker and Scarf (2020). Moreover, Tsokos et al. (2019a) provided details for a variety of Bradley-Terry and additional outcome-based models implemented in football data.

A home advantage can be incorporated into the model (Davidson and Beaver, 1977). The match outcome for the home team i competed against team j is represented by the random variable Y_{ij}, which takes the values zero, one, and two $(0, 1, 2)$ corresponding to a loss, draw or win, respectively. Under this perspective, the probabilities of the outcomes are defined by the equation

$$\Pr(Y_{ij} \leq k) = \frac{\exp\left(\delta_k + \eta + a_i - a_j\right)}{1 + \exp\left(\delta_k + \eta + a_i - a_j\right)}, \quad k \in \{0, 1, 2\}, \qquad (1.2)$$

where $\delta_0 < \delta_1 < \delta_2$ are cut-point parameters satisfying differentiability conditions, η is the common home advantage, and a_j is the ability of the team j—similar to parameters β_j of the original Bradley-Terry model; see Equation 1.1. In the above formulation, the sum-to-zero constraint $\sum a_j = 0$ is implemented for identifiability reasons. The above model is nothing more than an ordinal multinomial logistic regression model assuming proportional odds; see for details in Agresti (2013, Section 8.2.2).

The models described above measure the strength of each team j by using log-odds parameters, denoted by β_j for the BT model and by a_j for the ordered multinomial model specified by (1.2). In these models, the team abilities are treated as fixed effects over the entire season, implying that the variation in the match outcomes is attributed to other game-specific factors. However, this assumption of constant team strength is unrealistic and inconsistent with the views of sports experts and fans, especially for football. It is evident that teams experience different phases of performance throughout the season, and their ability parameters should reflect these changes.

Hence, these fixed ability models are extended as suggested by Cattelan et al. (2013) in order to allow for time-varying abilities. These dynamic ability models are extensions of the logistic and ordinal logistic regression models, which have been applied to sports data by Brillinger (2008, 2009). Goddard (2005a) employed an ordered probit regression model to forecast the outcome of a football match. In this model, the match result between teams i and j, denoted by R_{ij}, is determined by a latent variable y^*_{ij} and a Gaussian error term e_{ij}, which are assumed to be independent and identically distributed. The relationship between R_{ij} and y^*_{ij} is given by

$$R_{ij} = \begin{cases} 1 & \text{if } \delta_2 < y^*_{ij} + e_{ij} \\ 0.5 & \text{if } \delta_1 < y^*_{ij} + e_{ij} < \delta_2 \\ 0 & \text{if } y^*_{ij} + e_{ij} < \delta_1 \end{cases}$$

In the above equation, one stands for the home win, 0.5 for the draw and zero for the loss of the home team (or the win of the away team). The parameters δ_1 and δ_2 are thresholds that are estimated from the data. The latent variable y^*_{ij} can be modelled as a function of match-specific characteristics, whose coefficients are also estimated from the data. In the above specification, if we replace the Gaussian distribution of the error term with the logistic distribution, we obtain a multinomial logistic model—see Equation 1.2. A generalization of this model can be obtained by using a multinomial Dirichlet approach Diniz et al. (2019).

1.5 ELO type ranking

Given that paired comparison models are focusing on the analysis of ratings, it is reasonable use as predictors rankings or ratings generated by other models or mechanisms. For example, Hvattum and Arntzen (2010) used ELO ratings as covariates in a ordered logistic regression model. In football, this kind of rating system and its variations are widely used for the FIFA rankings[1] for national teams and the UEFA rankings[2] for club teams representing Europe.

The general idea is the following: each team i is assigned an initial rating represented by a real number X_i. When team i plays against team j, the ratings of both teams are updated by implementing a function $\Phi(\cdot)$ that depends on the difference between the current ratings, i.e. $\Phi(X_i - X_j)$. Note that the sum of all ratings remains constant; it is mathematically reasonable to centre these ratings in way that their sum equals zero—by this way, when a team wins some points, the other loses the same number of points. Practically, this rating scheme is adapted to the characteristics of each specific sport; for example in international football we have the FIFA/Coca-Cola World Ranking.

Ratings can be used for predictive purposes. Thus, we can answer questions like what is the probability of a team rated X to win a team rated Y? In tournaments which do not have a full round-robin format such as the World Cup or the Champions League, it becomes of primary importance to consider information which can make different groups comparable. Such information may be available via rankings which will leverage for the lack of games with all pair of teams competing to each other.

Finally, when we consider match outcomes in terms of home win/draw/away win, a variety of artificial intelligence and machine learning algorithms are available for direct implementation. However, this may come at the cost of reduced interpretability of the results and the potential risk of over-fitting in certain cases (Carpita et al., 2019; Tsokos et al., 2019a; Baboota and Kaur, 2019a).

1.6 Modelling the score: Issues to consider

An alternative and more detailed approach is based on modelling the score of the game. The term "score" we refer to the number of goals scored by each team and, therefore, it refers to a bivariate count. Alternatively one may consider just the goal difference between the two teams and hence work with an integer valued random variable defined in $\mathcal{Z}$.

1.6.1 Poisson or not Poisson

The selection of an appropriate statistical distribution is the first question which arises when modelling the final score in football. If pure chance

[1] `https://www.fifa.com/fifa-world-ranking/men`

[2] `https://www.uefa.com/nationalassociations/uefarankings/`

dominates the game, a Poisson distribution would be the obvious choice. However, football involves more than pure randomness. Each team possesses varying offensive capabilities, while defensive strategies and game-specific conditions can further influence the potential score.

Initial empirical investigations often reveal evidence of overdispersion in goal scored in a football match. This implies that the observed variance in goals scored in each match exceeds the expected variance under a Poisson assumption. This leads to the need for alternative models that can accommodate overdispersion, such as the negative binomial distribution. Additionally, relevant covariates, which can be considered as explanatory variables, can further alleviate this extra variability.

The presence of overdispersion has been acknowledged in the literature; see, for example, in Baxter and Stevenson (1988). While any mixed Poisson distribution can handle overdispersed data, the negative binomial distribution remains a popular choice due to its widespread use and established theoretical foundation. It is crucial to recognize that most models incorporating covariates estimate the mean of a Poisson distribution for each unique combination of covariates. This effectively utilizes multiple Poisson distributions rather than a single one, thereby mitigating the issue of overdispersion to some extent. In cases where observed overdispersion is relatively minor, this approach may be sufficient.

To illustrate this concept, consider Figure 1.1. Data from the English Premier League (EPL) spanning over 2000–2001 to 2020–2021 seasons. Each data point represents the mean and variance of goals scored by a specific team in a particular year. The diagonal line signifies equi-dispersion, while the dotted lines represent 95% confidence intervals for the variance given the mean. These confidence intervals depict the expected distribution under a Poisson model. As can be observed, a Poisson model appears reasonably adequate for describing the number of goals scored by individual teams within the EPL. It is important to note that due to the inherent variation in team strength, a single Poisson rate may not be sufficient to capture goal-scoring patterns across the entire league. Consequently, a slight degree of overdispersion might be expected.

While Figure 1.1 utilizes EPL data, similar trends are generally observed in other leagues. Conversely, international matches between national teams often exhibit higher level of overdispersion, potentially attributable to the greater differences in the strength of the teams at this level.

1.6.2 Correlated outcomes or not?

A second consideration regards the potential dependence between the goals scored by each team in the same match. Intuitively, when one team scores a goal, the opposing team is likely to increase its offensive efforts, potentially leading to a correlation between the number of goals scored by each side. This

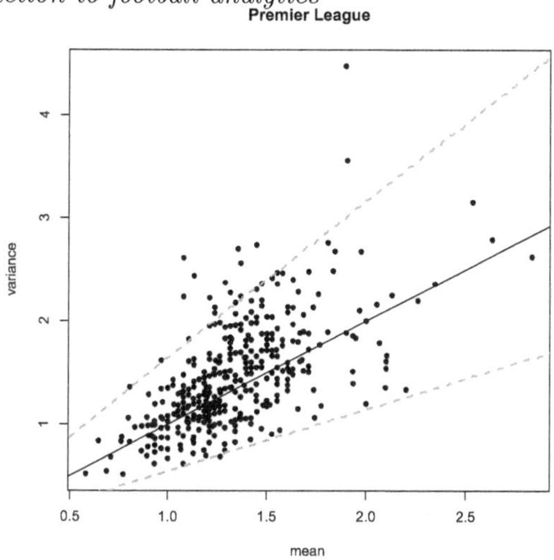

Each point refers to one team for a particular year. The diagonal line implies the equidispersion, while the dotted lines are 95% confidence intervals for the variance with given mean and imply what we would expect to see if the Poisson model was the true one

FIGURE 1.1
Mean and variance per team participation in the English Premier League (EPL) from 2000–2001 up to 2020–2021.

suggests the need for a joint model that considers the simultaneous occurrence of goals for both teams.

To examine and demonstrate this concept, we have used again the English Premier League (EPL) data from the 2000-2001 season onwards. The Pearson correlation coefficient was calculated for each championship season. Figure 1.2 presents the expected distribution of correlation coefficients under the assumption of no dependence (i.e., independent Poisson variables representing goals scored by each team in a simulated championship). The observed correlation values (red lines) fall within the expected range based on the simulation. While not definitively conclusive, this plot suggests that the observed correlation between goals scored by opposing teams is typically small.

While the above simple correlation analysis provides initial insights of potential dependence between goals scored by the two opposing teams in football, several crucial limitations must be considered. First, correlation analysis assumes a linear relationship between variables. This might be a suitable starting point, but the dependence between goals scored by each team might not necessarily be linear. Copula-based models, which will be discussed later, are more efficient at capturing non-linear relationships that could be present. Second, our observations suggest significant variation in the size and even the sign of correlation across different leagues. For instance, the Bundesliga exhibits negative correlations in most years, potentially reflecting distinct playing styles or defensive strategies compared to other leagues. This highlights the importance

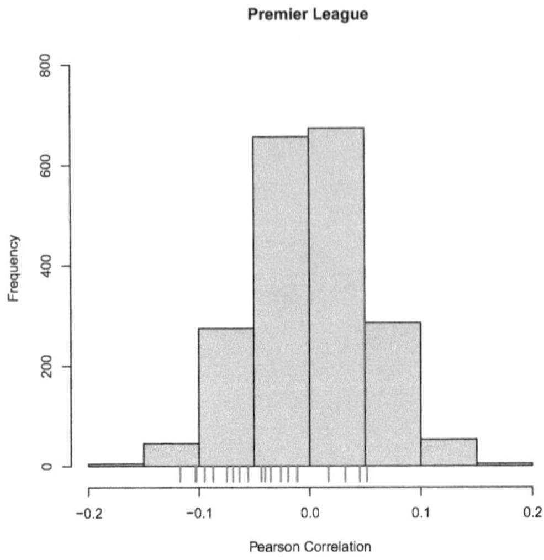

FIGURE 1.2
Distribution of the Pearson correlation assuming no correlation versus observed correlations for EPL data (fine vertical lines on x axis); The distribution is based on simulated data of 10,000 championship replications.

of considering league-specific factors when analyzing goal dependence. Third, the above naive analysis assumes teams with equal abilities, which is unrealistic. The relative strengths of opposing teams significantly influence goal-scoring patterns. A more sophisticated model should explicitly account for the variability in team abilities. Fourth, correlation behaves differently for discrete data, particularly for small counts like goal numbers. As highlighted by Karlis and Ntzoufras (2003), even minor correlations can significantly impact predicted win probabilities. This limitation necessitates careful interpretation of correlation values in the context of football score models. Fifth, the structure of competitions can also influence dependence. International matches, for example, might exhibit different dependence patterns compared to league matches due to factors like varying team strengths and less frequent matchups between teams. Acknowledging these limitations and exploring more sophisticated techniques like copula-based models becomes crucial to accurately model the dependence between goals scored by opposing teams in football.

In conclusion, the aforementioned considerations are critical factors in selecting an appropriate model for predicting football game scores. Moreover, the heterogeneity of data across leagues highlights the need for specialized modelling approaches, potentially contributing to the development and implementation of different models in the field.

1.6.3 Which covariates to use?

Another question concerns the type of covariate information necessary for the modelling approach. The selection of covariates for sports modelling is highly dependent on the modelling purpose. Descriptive models, aiming to explain past outcomes, usually require different covariates compared to predictive models that aim to forecast future results.

One naturally considered covariate for descriptive models is home advantage. This well-established phenomenon in various sports, including football, refers to the tendency of teams to perform better at home. Home advantage has been attributed to factors such as fan support, familiarity with the environment, and travel disadvantages for away teams.

Predictive models can benefit from additional covariates beyond home advantage. These may include team composition, player injuries, and weather conditions. The specific impact of these variables might depend on the particular teams involved and the nature of the match. Incorporating information from betting agencies can be also valuable for predictive models. Betting odds reflect expert opinions about the outcome and may contain valuable insights.

The inclusion of specific covariates often depends on the championship under investigation. While various features and factors have been proposed as potential covariates in the literature, the statistical significance of some of them still remains questionable. Additionally, exploratory models, particularly for international tournaments like the World Cup or European Championship, may incorporate socioeconomic variables (Groll and Abedieh, 2013; Groll et al., 2018a).

1.6.4 Temporal correlation or constant across time?

An important consideration for modelling sports data is the potential presence of temporal dependence. Since teams play matches sequentially (usually week-by-week arranged in fixtures or match days), observations may not be independent across time. This raises the question of whether the data should be treated as a time series.

Several studies have addressed this issue. For example, Harvey and Fernandes (1989) investigated the time series characteristics of the number of goals scored by England against Scotland in their annual match (known as "England Scotland football rivalry") Their model explicitly accounted for the temporal characteristics of the data. Similarly, a recent study by (Mattera, 2023) employed a binary time series analysis to forecast outcomes. Angelini and De Angelis (2017) implemented a count time series model to capture temporal correlations in their analysis.

Furthermore, the effects of covariates, particularly those related to team strength, can also exhibit temporal dependence. Factors such as training schemes, participation in other tournaments, and player injuries can influence a team's performance and may change over time. Teams' performance is typically dynamic, potentially fluctuating across seasons or even weeks. This dynamism can be attributed to various factors, including roster changes, player fatigue, coaching changes, motivational shifts, and cyclical effects due to training programs. To account for this temporal dependence in team performance, researchers have developed dynamic models that allow certain effects to vary over time, resulting in greater model flexibility.

1.7 Models, models, models ...

In this Section we introduce and briefly discuss some further models that are more specialized to treat specific sports problems or certain aspects of a specific sport.

1.7.1 Basic models for the number of goals

In football, scoring goals is the ultimate objective, making it a natural choice for the outcome underthe goal-based models. Compared to win/draw/loss models, goal-based models offer additional benefits, such as enabling the calculation of implied team abilities.

The initial approach often involves considering univariate discrete distributions for the number of goals scored by each team. Under the simplistic assumption that randomness solely dictates football outcomes, Poisson distributions could be used to model goal counts. However, this approach disregards the influence of factors beyond randomness, such as team skill and playing styles.

A critical question then arises: Are alternative distributions more appropriate to account for the inherent variability or over-dispersion in goal scoring? Additionally, since the two teams compete against each other, incorporating a correlation structure seems to be essential for modelling the final score in a realistic way.

Let X and Y represent random variables denoting the number of goals scored by the home team and the away team, respectively. The final match outcome can be categorized as a win for the home team if $X > Y$, a draw if $X = Y$, and a loss if $X < Y$. Given an appropriate statistical, model we can calculate the probabilities of these outcomes: $\mathrm{P}(X > Y)$, $\mathrm{P}(X = Y)$, and $\mathrm{P}(X < Y)$. To achieve this, we require a probability model for the joint distribution of (X, Y). This model can be based on either a discrete distribution assuming independence (where the joint probability is the product of the marginal probabilities) or a model incorporating dependence, which necessitates specifying a specific joint discrete distribution. The current literature offers a rich landscape of statistical models for modelling or predicting football scores. A detailed description and discussion of these models will be presented in Chapters 4 and 5. However, we briefly introduce the variety of existing approaches here to provide a preliminary overview.

As previously mentioned, a common starting point for modelling the number of goals scored by each team is the Poisson distribution. For joint modelling, one approach leverages two independent Poisson distributions, known as the *double Poisson model* (Lee, 1997). This, model when combined with a covariate structure, can provide a good fit in many applications. An alternative approach, proposed earlier by Maher (1982), utilizes a correlated Poisson bivariate distribution. This model has been employed in specific contexts, and more recently, it has been integrated with advanced variable selection techniques, as demonstrated by Groll et al. (2018a).

The next step often involves refining the above initial models to better capture specific characteristics observed in the data. For instance, Dixon and Coles (1997) proposed an extension of the double Poisson model by adjusting the probabilities of particular scores, especially (0,0), (0,1), (1,0), and (1,1). This modified model has gained an increased attention and applied in several different occasions. Karlis and Ntzoufras (2003) introduced another approach that increases the probabilities across the entire diagonal of the joint goal distribution which refers to draws. This effectively allows for a higher probability of each draw. The observed excess of draws was more common when the points awarded for win-draw-loss followed the 2-1-0 system. This has decreased, and in some cases, and in some cases diminished, with the adoption of the more modern 3-1-0 point system.

To account for more flexible correlation structures between the number of goals scored by each team, McHale and Scarf (2007, 2011a) proposed a copula-based model. Copulas allow researchers to specify independent marginal distributions (for example Poisson) and then introduce dependence through a chosen copula function. Boshnakov et al. (2017) explored a different approach. Recognizing that the time between goals scored follows an exponential distribution in Poisson models, they derived a model based on the Weibull distribution. This leads to a discrete Weibull distribution for the number of goals, coupled with a copula to account for correlation and propose a new bivariate model.

A common feature across these models is the incorporation of team-specific strength parameters and outcomes within a probabilistic framework. Additionally, they all account for the home advantage effect. A widely used method for modelling team attack and defence parameters within a Poisson framework is often attributed to Maher (1982). This approach remains a popular choice, as evidenced by its continued use in the recent works by Koopman and Lit (2015) and Koopman and Lit (2019a).

1.7.2 Dynamic models

The models discussed in Section 1.7.1 assume team ability parameters that remain constant over time. However, in practice, dynamic models that allow team-ability parameters to evolve across time are more realistic and can be more effective in practice although they introduce increased mathematical complexity. Rue and Salvesen (2000) proposed such a model based on a Bayesian dynamic generalized linear model, while Owen (2011) introduced a similar approach for a bivariate Poisson model. More recent work by Koopman and Lit (2015) also explores dynamic extensions. Additionally, the work of Crowder et al. (2002) presents another relevant model in this area.

1.7.3 Models for the goal-difference

An alternative approach is to focus on modelling the goal difference (also known as the margin of victory) between the two opponent teams rather than the goals scored by each team. This approach offers several advantages. Firstly, it eliminates the inherent correlation arising from the fact that the two teams compete against each other (Karlis and Ntzoufras, 2009). Secondly, it avoids the assumption of marginal Poisson distributions for the individual team goal counts, allowing for greater flexibility (Shahtahmassebi and Moyeed, 2016). Additionally, it aligns with certain betting markets, such as the Asian handicap, which focus solely on the goal difference. A clear disadvantage of this approach is that such models use less match related information compared to models that directly model individual team scores. For example, we will not be able to infer for the total number of goals scored in each game which is of prime interest in some bets. However, they still offer more information than basic outcome prediction models discussed previously.

Karlis and Ntzoufras (2009) proposed a model based on the Skellam distribution which is the distribution derived from the difference between two independent Poisson random variables. However, the Skellam distribution can also be derived under specific other conditions. Shahtahmassebi and Moyeed (2016) proposed to use a similar model based on the distribution of the difference of two generalized Poisson random variables. Furthermore, Manderson et al. (2018) extended this approach to introduce a dynamic model modelling the difference of the goals. Gaussian distributions were used in earlier attempts to model the goal difference by Stefani (1983) and Heuer and Rubner (2009). However, this approach has severe limitations compared to the methods discussed above since the normal distribution is not appropriate mathematically for modelling a discrete random variable.

1.7.4 In-play models

The models discussed so far have focused on analysing and predicting the final outcome of a match. However, there is growing interest in the world of betting for models that can predict outcomes during the course of the game, also known as in-play modelling. In-play modelling aims to predict the final outcome of the match based on information available during the game. Examples include calculating the probability of a team winning if the score is 1-0 at the 20-minute time point, or the impact of a red card on the final result. A crucial question for such models is which type of in-play information can improve prediction accuracy. While some existing models can be adapted to provide conditional probabilities, incorporating additional in-game information often necessitates modifications of such models. Dobson et al. (2017) and Asif and McHale (2016) provide examples of such models.

Titman et al. (2015) employed a real-time, eight-dimensional multivariate counting process to analyze the interplay between various events within a football match. This approach not only modelled the interdependence between home and away team goals, but also sought to quantify the influence of cards on the game's outcome. The findings of their study suggested that yellow cards did not significantly impact goal scoring rates, whereas red cards, particularly when issued to the away team, had a substantial negative effect. Recent advancements incorporate survival analysis techniques (Zou et al., 2023), while Klemp et al. (2021) proposed using event data observed within a specific match to predict the final match outcome.

1.7.5 Survival analysis models

In this section, we will discuss about models that do not model the number of goals, but rather focus on modelling the time intervals between goals. Nevertheless, the two approaches are directly connected since, for example, assuming a Poisson distribution for the number of goals scored by a team implies that the time between consecutive goals can be modelled using an exponential distribution.

Modelling the time intervals between goals offers several advantages, particularly regarding how we incorporate information during the match. This approach in our analysis. Therefore, it can be valuable for in-play models prediction. Several relevant research works have explored this topic, including Volf (2009), Dixon and Robinson (1998) and Nevo and Ritov (2013). Another related work is the one by Thomas (2007) who investigated the inter-arrival times of goals in ice hockey using Weibull and Plateau-Hazard distributions. As suggested by Boshnakov et al. (2017), the assumption of Weibull-distribution for the inter-arrival times is reasonable.

However, compared to research on goal counts, significantly less work has been dedicated to modelling the goal arrival times.

1.7.6 More information about predictive models

For a comprehensive review of existing predictive models, we refer readers to the works of Scarf and Rangel Jr (2017) and Pearson et al. (2020). A more detailed description of the models discussed here is provided in Chapters 4 and 5.

1.8 Other modelling aspects

Beyond game outcome prediction, models for football data can serve another crucial purpose: to investigate broader hypotheses about the game itself. In this context, the focus shifts away from predicting the final scores and instead may centre on understanding and analyzing specific game characteristics and their potential to influence on the final match outcome. This research direction has proven highly fruitful, yielding numerous interesting findings in the literature. Consequently, the need for accurate and well-specified models becomes very important. To illustrate, some potential areas of exploration using football models include (but are not limited to):

- The home effect advantage;
- The effect of altitude and of the artificial pitch;
- The red (and yellow) card effect;
- The contribution of Video Assisted Referee (VAR) technology;
- The use of tracking data;
- The optimal time for substitutions;
- Estimation and effect of competitive balance.

1.8.1 The home advantage

Home advantage, a well-established phenomenon in sports, has been the subject of considerable research in football as well. The concept was first examined within the context of the World Cup by Dowie (1982), who observed a success bias for host countries. Building upon this foundation, Pollard (1986), with their seminar paper, laid the basis for a more comprehensive investigation of this effect. Using data from various English and European competitions, Pollard explored the influence of crowd support, travel fatigue, team familiarity, potential referee bias, tactical adjustments, and psychological factors on home advantage.

Subsequent studies have further studied specific aspects of home advantage. These investigations have examined the impact of pitch surface (Barnett and Hilditch, 1993), travel distance (Clarke and Norman, 1995; Nevill et al., 1996), potential referee bias (Nevill et al., 2002), and territoriality (Neave and Wolfson, 2003).

More recently, with the emergence of COVID-19, there has been significant interest in the impact of playing behind closed doors on home advantage. Several studies, including those by Benz and Lopez (2023), Fischer and Haucap

(2021), and Sors et al. (2021), have explored how the absence of spectators may have affected the final results.

A special case of the home effect is the **effect of altitude** on (certain) stadiums, which may provide the home team a competitive advantage. For example, national teams like Bolivia and Chile strategically try to exploit their high-altitude of their home stadiums, creating a potential confounding factor with the previously discussed (general) home advantage effect.

This issue sparked the "*high-altitude football controversy*" when FIFA, in 2007, banned World Cup qualifiers from being played in stadiums exceeding 2500 meters above sea level. This decision impacted Bolivia, Ecuador, and Colombia, preventing them from hosting qualifiers in their capital cities. The limit was subsequently raised to 3000 meters, ultimately affecting only Bolivia. However, the ban was lifted in May 2008.

Despite the controversy about the best policy and the eventual withdrawal of any altitude-related ban, the question about the impact of stadium altitude on match outcomes persists. Chumacero (2009) examined this using a bivariate Poisson model for international match outcomes but found no significant effect. However, Casas and Fawaz (2016) reported evidence of an altitude effect for some South American national teams using a different dataset. Therefore, there is no clear conclusion about the effect of stadium altitude, and is still an open issue of dispute in the academic community.

Similarly to the influence of altitude, another issue that can contribute to the home effect is the use of **artificial pitch surface**. Concerns about natural grass maintenance, particularly watering requirements in hot climates, have led to the increased adoption of artificial grass pitches. This trend is particularly evident in some countries with dry climates. While traditional football has been played on natural grass surfaces, there is a growing shift towards artificial surfaces at all levels of the game. Major League Soccer (MLS), the highest professional league in the United States and Canada, is an major case example. In the 2014 season, four out of nineteen teams played their home matches on artificial grass (AG).

Despite this rise in popularity, artificial grass pitches continue to generate controversy among players and coaches. There are at least three primary reasons for this resistance. The first concern relates to potential player injuries. While research findings regarding injury risk on artificial versus natural grass remain inconclusive, some high-profile MLS players, including David Beckham and Thierry Henry, have expressed concerns and even refused to play on artificial surfaces. The second reservation is the perception of increased fatigue experienced by players on artificial grass. Finally, some players believe that the ball behaves differently on artificial surfaces, traveling faster and bouncing higher, potentially creating an advantage for certain teams. However, research by Barnett and Hilditch (1993) and Trombley (2016) did not find significant evidence to support these claims.

1.8.2 Card modelling and the red card effect

What about the effect of red cards in football? Red cards are an essential characteristic of football. The are disciplinary measures where a player is expelled (without being substituted) for severe rule violations or dangerous play. A red card may severely influence the outcome of a football match. It necessitates immediate tactical adjustments, primarily for the team with the player disadvantage (and secondarily for the opponent). Spectators, coaches, and bookmakers alike are interested to quantify the magnitude of this effect: how dramatically can a red card change the strength balance between the two opponent teams?

This topic has been thoroughly investigated in studies by Ridder et al. (1994) and Mechtel et al. (2011). The challenge lies in the dynamic nature of this effect, as the time point that the red card occurs and the momentum of the match appears to be of primary importance that interact with the red card's influence.

Based on economic theory, these studies analyze match data from the German Bundesliga (1999-2009). Their findings reveal a negative performance impact for home teams that receive red cards. For away teams, however, the effect is contradictive and depends on the remaining game time. Interestingly, the results suggest that a late red card for the away team can even be advantageous. This potentially suggests that the "ten do it better" myth holds some truth for guest teams to some extent.

Dawson et al. (2007) implemented various models with response the occurrence of red cards. These models were similar to those used for goals prediction. For instance, Dawson et al. (2007) explored the use of a double negative binomial model using Frank a copula to account for correlation. He also developed a zero-inflated version of this model. Both of these models were an initial attempt to quantify the potential influence of referees on card implementation. Thus, a related area of investigation concerns the potential referee bias in awarding disciplinary sanctions. Buraimo et al. (2010) examine this issue in the context of the English Premier League and the German Bundesliga. Their research explores potential biases in referees' decisions regarding warnings (implied by yellow cards) and player expels (red cards) by incorporating in-game information within match-level models.

1.8.3 The contribution of the video assisted referee (VAR)

The recent introduction of the video assistant referee (VAR) system has generated discussion within the football world. Implemented in most major championships, including the World Cup since 2018, VAR aims to improve the accuracy of referee decisions through video review of potentially game-changing events which are unclear by a simple visual inspection.

Spitz et al. (2021) investigated the impact of VAR by examining 2195 matches across 13 countries. Their findings reveal that VAR conducted 9732 checks for unclear events, with a median duration of 22 seconds per check. The study demonstrates that the odds of a correct decision were significantly higher after VAR intervention compared to the initial referee's decision.

Further insights into the effects of VAR can be found in the works of Carlos et al. (2019) and Lago-Peñas et al. (2021). These studies explore the impact of VAR on specific game characteristics by analysing data from various leagues before and after its implementation. For example, Carlos et al. (2019) examine potential changes in gameplay, while Lago-Peñas et al. (2021) focus on the Spanish La Liga.

1.8.4 The use of tracking data

In recent years, player tracking technology was introduced in football mainly in more advanced competitions and in wealthier teams. This technology has allowed these teams immediate access to information of player movements across the whole pitch, throughout a match. These devices generate massive datasets throughout a match, capturing player movements at a frequency of 30 observations per second. This detailed information allows for in-depth analysis of player attributes and behaviours, providing valuable insights into team tactics, passing networks, space creation, and overall coordination. Therefore, the use of tracking technology and obtained data, according to experts, has revolutionized football analysis.

The most popular and comprehensive metric resulted by such technology are the "*expected goals*" (xG). This metric measures the quality of a scoring opportunity by considering factors such as short distance, angle, and type (header, free kick, etc.). This is achieved via the implementation of a logistic regression model which estimates the probability of success of each chance or shot. By comparing xG to actual goals scored, analysts can assess the offensive and defensive efficiency of each team. However, xG remains a work in progress. It is open to critique by sport analysts since the existing models do not account for defensive actions, player skill variations, and its generalizability across different leagues.

For further details on tracking data analysis, we refer the interested reader to the work of Goes et al. (2021). Their comprehensive survey of the football tracking data literature explores existing research and identifies promising areas for future investigation.

1.8.5 Planning the optimal time for substitutions

The question about the best time point for substitutes in football has returned in the foreground after the recent increase of the number of substitutions from three to five in most competitions (excluding the English Premier League). This adjustment, combined with an extremely busy fixture calendar, has highlighted the need of maintaining player fitness and weariness.

Myers (2012) proposed a specific substitution plan for teams that are behind in score during a match. They suggested that making the first substitution by the 58th minute, the second by the 73rd minute, and the third by the 79th minute can potentially double the chances of a comeback in the match. However, the authors acknowledge that even with this strategy, the odds of successfully reversing a match outcome remain less than one.

Silva and Swartz (2016) offer a different perspective on substitution timing. The authors review the substitution rule proposed by Myers (2012) and provide a discussion of the results. They further present a new approach based on Bayesian logistic regression. According to their findings, they did not identified any specific time point where substitutions offered a clear advantage.

1.8.6 Competitive balance: A key factor in fan engagement

Competitive balance, in the context of sports leagues, refers to the closeness of playing strengths of participating teams. This concept is closely associated to the inherent uncertainty of sporting outcomes, where the final result is not predetermined. This uncertainty is that makes football so captivating for fans. Football leagues or competitions with a high degree of competitive balance tend to attract larger audiences due to the unpredictability of match results, stimulating excitement and interest in the sport.

Competitive balance has become a central concept within the economic theory of professional sports leagues. Growing recognition of its multifaceted impact has generated further research. For instance, a strong correlation exists between competitive balance and fan attendance/welfare. Fans are captivated by matches that offer an increased level of unpredictability, as it enhances the thrill and excitement of the competition; for a recent review on this topic, see in Pawlowski and Nalbantis (2019).

Finally, football prediction models can serve as a basis to calculate competitive balance indices (see, for example, in Deb, 2022). Ongoing research efforts are directed towards quantifying and understanding the uncertainty that remains unexplained by such models.

1.8.7 Concluding thoughts and discussion

While this section highlights several modelling topics of ongoing research in football, it is by no means exhaustive. Research in this area continues to evolve, tackling not only new modelling challenges and specific questions but also interesting methodological innovations.

One such challenge involves comparing different models and assessing their ability to predict outcomes or scores. The question of how to measure model success has been addressed in the literature, with various metrics proposed such as the Brier score and the ranked probability score. Constantinou and Fenton (2012a) provide a more in-depth discussion on this topic, which will be further explored in Chapters 3 and 7.

1.9 Organization of the book

As discussed in this chapter, football modelling has emerged as a flourishing research area that offers a powerful quantitative tool for understanding and analyzing the game of soccer. By analysing data and models effectively, all football related stakeholders, from teams to spectators, can be the receivers of valuable insights. The remainder of this book is organized as follows.

Chapter 2 introduces and presents the necessary tools how to organize data from football and how to implement some basic models and extract information from them. This chapter demonstrates the application of a simple model for prediction purposes, exploring various estimation methods, including those based on maximum likelihood and Bayesian approaches.

Chapter 3 discusses and illustrates league table predictions using Monte Carlo simulations. It also covers metrics for evaluating model effectiveness and for performing model comparisons. Finally, the chapter addresses model checking procedures to ensure the validity of model assumptions and criteria.

Chapter 4 goes deeper into several basic existing football prediction and modelling approaches, providing comprehensive mathematical details. We present these models in a self-contained manner, offering sufficient information on each model and underlying the connections between them. The chapter follows an incremental structure, building upon previously introduced models before presenting new ones. It begins by exploring some fundamental models, starting with the double Poisson model, and progresses through detailed explanations of the bivariate Poisson model and their dynamic extensions. For each model, we will discuss their formulation, estimation methods (briefly), potential limitations, unique contributions, advantages, disadvantages, and their implementation through the footBayes R package accompanying the book. We will provide a detailed description of the package's functionalities, accompanied by extensive illustrative examples. We will in fact deeply introduce this package in order to fit the basic (and more advanced) models and extract the main summaries by using the Italian Serie A 2009/2010 data as a motivating example.

Next chapter, Chapter 4, focuses on additional models, such as those designed to model the goal difference—Skellam and student-t models. The practical implementation of these models will be illustrated through some R packages, mainly the aforementioned footBayes package. This chapter introduces the essential mathematical concepts underlying the models and examines their application to the Italian Serie A dataset.

An illustrative modelling experience for the Euro 2020 and World Cup 2022 tournaments is provided in Chapter 6. The analysis and the predictions are obtained through the footBayes package described in Chapters 4 and 5.

The final chapter, Chapter 7, explores the intersection of football modelling and betting. The betting industry is a major consumer of football data and also generates valuable data from its own models. This chapter introduces some fundamental concepts related to betting, including the relationship between odds and game probabilities, using examples for clarity. We will also discuss betting strategies, such as the well-known Kelly criterion. Our aim is to highlight how models can play a crucial role in, betting even though odds are also influenced by market factors. By considering betting odds as implied probabilities, we will discuss their usage, comparison methods, and practical applications for bettors, while acknowledging the additional information that bettors themself contribute to the market. We will provide a practical case study to construct some betting strategies in order to win money from the bookmakers betting companies.

The book is accompanied by an appendix which serves as a reference for more advanced mathematical and computational details, particularly for specific methodologies and models.

2

Methods, algorithms and computational tools

2.1 Model formulation

In football analytics models for the final goal score, we consider a dataset where every observation $i \in \{1, \ldots, n\}$ in our dataset (row in the associated data file) will refer to each game/match; n is the number of games. As responses we consider the number of goals scored by each team denoted by Y_{i1} and Y_{i2} for the first (usually the home team) and the second (usually the guest or away) team. The rest of the columns in our dataset will record the two competing teams and specific characteristics of the game or the teams playing the game. In many occasions the explanatory variables, or features, are usually performance indicators for the two opponent teams. This data format will be refereed at the bivariate-data format.

We can have two different types of models: (a) a predictive model, or (b) a descriptive or interpretable model. The aim in a predictive model is to be able to efficiently predict the final score of the game with information which is available before the beginning of the game. On the other hand, a descriptive or interpretable football analytics model aims at understanding the game itself and what makes a team a winner. The latter models are more useful for football managers since they help them to understand what are the weaknesses and the strengths of each team. By this way, football managers can make correction moves in practice or in transfers to improve the performance of their team.

As it is obvious the covariates we consider in these two types of models are different. The predictive models can use information and team performance indicators from the "near" past while the interpretable models can use the performance metrics of the current game (usually called box-score statistics) that are only available after the end of the game.

Hence, generally we can write that

$$(Y_{i1}, Y_{i2}) \sim f(\boldsymbol{\theta}_{i1}, \boldsymbol{\theta}_{i2}, \boldsymbol{\rho}_i) \quad \text{for } i \in \{1, \ldots, n\}, \tag{2.1}$$

where $\boldsymbol{\theta}_{i1}$, $\boldsymbol{\theta}_{i2}$ are parameter vectors referring to the distribution of the goals of each team, while $\boldsymbol{\rho}_i$ are parameters referring to the joint modelling of the two responses (including correlation or equivalent association parameters).

DOI: 10.1201/9781003186496-2

2.1.1 The double Poisson model

The simplest form of (2.1) is the so-called double Poisson (DP) model where we assume

$$Y_{i\ell} \sim \mathcal{P}oisson(\lambda_{i\ell}), \quad \text{for } \ell = 1, 2 \text{ and } i \in \{1, \dots, n\}. \tag{2.2}$$

This model can be easily fitted within the framework of the Generalized Linear Models (GLMs) (Nelder and Wedderburn, 1972) using standard software such as R or SPSS. We only need to re-arrange our data such that each game will take two rows in the dataset (with a total of $2n$ rows) and one response which will be the number of goals scored by each team and with the covariates repeated in an appropriate way—see Section 2.1.2 for a more detailed description.

Moreover, the above model assumes independence between the goals scored by the home and away team conditionally on the model parameters. Any game association or correlation will be indirectly introduced via the estimated model parameters.

This simple model will be considered here as the basic springboard in order to extend our modelling approach and build more realistic models which will account for the unique, specific characteristics of association football. Note that the Poisson distribution plays a central role in football score models (see Chapter 5) and it is commonly used to describe the number of successes (here number of scored goals) within a fixed time interval (here usually 90 minutes + added time). Poisson-based models are often also called Poisson regression or Poisson log-linear models; the latter name has its source in the use of the logarithmic link within the GLMs framework, being the logarithm the standard canonical link function in such models. Generally we can write

$$\log \lambda_{i\ell} = \eta_{i\ell}, \text{ with } \eta_{i\ell} = \psi(X_{i1}^{(1)}, \dots, X_{ip}^{(1)}, X_{i1}^{(2)}, \dots, X_{ip}^{(2)})$$

where $\eta_{i\ell}$ is the predictor for game i and team ℓ ($\ell = 1$ for the home team and $\ell = 2$ for the away team) when using the bivariate data arrangement. The predictor $\eta_{i\ell}$ is a function of the covariates $X_{ij}^{(1)}$ and $X_{ij}^{(2)}$ which denote the j-th feature/covariate values of the i-th game for the home and the away team, respectively. Usually, we consider the simple function for the predictor, i.e. we consider a linear predictor which takes the form

$$\begin{aligned} \eta_{i1} &= \beta_0 + \sum_{j=1}^{p} \beta_j^{(1)} X_{ij}^{(1)} + \sum_{j=1}^{p} \beta_j^{(2)} X_{ij}^{(2)}, \\ \eta_{i2} &= \beta_0 + \sum_{j=1}^{p} \beta_j^{(1)} X_{ij}^{(2)} + \sum_{j=1}^{p} \beta_j^{(2)} X_{ij}^{(1)}. \end{aligned}$$

In the following of this section we will refer to different covariate structures used for the (linear) predictor of such models. We will start from the simplest model which is called the "vanilla" model and has been established through the years as the easiest and basic starting model not only for football but also for other team sports.

2.1.2 The vanilla model structure

The simple vanilla model has its origin back to 1982 in the original work of Maher (1982). It was subsequently used by other researchers like Lee (1997) and Karlis and Ntzoufras (2000a) to analyze football data. Following these early publications, Karlis and Ntzoufras (2003) and other researchers, such as Baio and Blangiardo (2010), Egidi et al. (2018b) and Owen (2011), have used this formulation in order to build efficient extensions.

The main characteristic of this model is that it is very simple and needs minimal information in order to be fitted. When a reasonable amount of games is gathered, then its accuracy is surprisingly adequate given the minimal information used. Hence any new formulation should be compared with such a basic model.

Using the bivariate-data formulation, the double Poisson vanilla model is written as

$$\begin{aligned} \log \lambda_{i1} &= \mu + home + att_{h_i} + def_{a_i} \\ \log \lambda_{i2} &= \mu \qquad\qquad\;\, + att_{a_i} + def_{h_i}, \end{aligned} \tag{2.3}$$

for $i = 1, \ldots, n$. In the above model formulation:

- n is the number of games under consideration;
- μ is a constant parameter;
- *home* is the home-effect;
- h_i and a_i are home and away teams in game i;
- att_k and def_k are the attacking and defensive effects or "abilities" of the k-th team for $k = 1, 2, ..., K$; and
- K is the number of teams in the data; usually $K \in 14, 16, 18, 20$ for major national domestic leagues.

This model simply implies that the number of goals scored by each team depends on the home-effect benefit (which is well established in the literature), the attacking ability of the scoring team, and the defensive ability of the team accepting the goals. Note that the effect of the attacking abilities is positive, which means that the greater is the team ability, more goals are scored by this team and hence its performance is better. On the other hand, the effect of the defensive abilities is negative on the log-expected goals of the scoring team. Hence, large negative values will decrease the expected number of goals of its opponent. Therefore, higher defensive ability values indicate teams with worse defensive ability, while smaller values imply teams with greater defensive power.

Interpretation of model parameters

In order to make the above model identifiable, we use the sum-to-zero constraints on the team ability parameters att_k and def_k. Hence, we impose the constraints

$$\sum_{k=1}^{K} att_k = \sum_{k=1}^{K} def_k = 0.$$

The above constraint in practice implies that in the estimation procedure we estimate $K - 1$ parameters while the missing parameter is simply estimated by the equations

$$att_1 = -\sum_{k=2}^{K} att_k \text{ and } def_1 = -\sum_{k=2}^{K} def_k \; .$$

In the above equations we have removed the first parameter corresponding to the ability parameter of the first team. Another usual choice is the last parameter (denoted here by K). The values of the estimated parameters will not be affected by the choice of the missing parameter. In the following, we will assume that the first parameter is eliminated from the model formulation and it is calculated as a simple function of the remaining ones.

Note that the sum-to-zero (STZ) constraint is preferable here over the more usual corner (or treatment) constraint or parametrization for interpretation reasons. The STZ approach will produce abilities which are compared with the overall team scoring ability (in the log-scale here). Hence positive attacking ability parameters mean that the team is better than an average team while negative values imply that the attacking team ability is below average.

Similar is the interpretation for the defensive abilities but with opposite sign, since higher value here implies worse defence or restraining power. Hence positive defensive parameters indicate teams with defensive ability worse than an average team while negative defensive parameters indicate teams with defensive ability better than an average team. For more details about these constraints, we refer the reader to Chapters 4 and 5.

Finally, the exponent of the constant parameter (e^{μ}) provides the expected number of goals of the away team in a game between two teams of average attacking and defensive strength. Equivalently, the value of $e^{\mu+home}$ provides the expected number of goals of the home team in a game between two teams of average attacking and defensive strength. Therefore, the exponent of the *home* parameter simply provides the relative increase of the expected home goals in a game between two teams of average attacking and defensive strength. The latter can be further used also for two teams of equal strength (where $att_{h_i} = att_{a_i}$ and $def_{h_i} = def_{a_i}$) since

$$\begin{aligned} \log\left(\frac{\lambda_{i1}}{\lambda_{i2}}\right) &= \mu + home + att_{h_i} + def_{a_i} - \mu - att_{a_i} - def_{h_i} \\ &= home + (att_{h_i} - att_{a_i}) + (def_{a_i} - def_{h_i}). \end{aligned}$$

Furthermore, if we consider the log-ratio of the expected goals in each game then we obtain

$$\begin{aligned} \log\left(\frac{\lambda_{i1}}{\lambda_{i2}}\right) &= home + (att_{h_i} - def_{h_i}) - (att_{a_i} - def_{a_i}) \\ &= home + ability_{h_i} - ability_{a_i} \end{aligned} \tag{2.4}$$

where $ability_k = att_k - def_k$ can be considered as overall team abilities which are simply given by the difference between each team's attacking and defensive ability. Returning to the interpretation of the home effect, e^{home} can be generally considered as the relative increase of the expected home goals in a game between two teams of equal strength (or overall abilities).

2.1.3 Additional features for prediction or interpretation of the game

As we have already discussed, the vanilla model is an initial starting point in order to build more sophisticated predictive or descriptive models for association football.

The types of covariates/features that can be used to enhance the model formulation depends on the purpose of the analysis. If prediction is the aim, then performance metrics based on previous games (usually averages) or older historical data can be used (Ulmer et al., 2013; Tsokos et al., 2019a). Moreover, economic data such as overall budget and transfers are also of prominent importance (Egidi et al., 2018b). On the other hand, if "understanding the game" in order to help the manager or the team officials to improve the team is the aim, then box-score statistics based on each and that are only available at the end of the game can be used instead. In both occasions, but more often in the predictive modelling approach, the number of covariates p which can be collected is large, usually larger than the number of the observed games n. Hence, shrinkage methods such as lasso (Tibshirani, 1996) are used to get-rid-off fast and efficiently variables that are not useful in the final model (Groll et al., 2015).

In the following we will call such football models based on the use of team features/covariates as (predictive or descriptive) performance models in order to discriminate them from the simple vanilla structured models. This label is used conventionally due to the prominent importance of the team performance metrics in determining the final score of a football game and it does not mean that any additional team non-performance features (such as economic indicators) cannot be included in the model formulation.

2.1.4 Performance features vs team abilities

When using performance-based models (discussed in Section 2.1.3), the team abilities of the vanilla model may become non-significant. The main reason for this, is that the ability parameters usually carry similar information as the ones extracted by performance metrics of other team characteristics. This is one of the main reasons why performance models with an increased number of covariates might not demonstrate, as much as we would expect, improved model fit or predictive ability; see in Van Eetvelde et al. (2021) for an illustration where the two types of models converge on similar predictive ability when the number of games increases. Similar results were reported in Tzai et al. (2021) for Basketball games in the Greek and the Spanish national leagues.

Since the two types of model will provide similar quality of predictions, a reasonable question is why to use predictive models which require a large amount of additional covariate information instead of the much simpler vanilla models. The main reason for using performance-based models is that vanilla models need an increased number of games in order to be able to estimate the performance of each team. Moreover, vanilla models work efficiently in full balanced leagues of round-robin type of tournaments and not in elimination (knock-out) cups or hybrid tournaments. Especially, in the latter format, when a group-stage of several mini-round robin leagues is followed by elimination phases, the vanilla model will totally fail unless we include some data from the knock-out phase. The reason is that the abilities of each group (mini-round robin league) are calculated relatively to the teams within each group. So these abilities cannot be used to compare teams of different groups. After at least one knock-out phase, we will have information across the abilities of the teams of different groups since we will have some cross-over games. Nevertheless, this information will be again weak since it will be based only on a few games. Finally, the abilities of each team need data only of the current season. Hence, at the beginning of each season, it is not reliable to use a vanilla model for prediction. Note that data of the first half of a round-robin league are enough in order to generate reliable predictions using the vanilla model (even for the final rankings of the league).

On the other hand, the performance models are based on the overall association between general performance indicators (or other characteristics) with the final score. Assuming that the game changes slowly across time, data from previous seasons are relevant, so we can use a performance-based model also for prediction after a few games. For hybrid tournaments such as Euro, World cups and Champions' League, overall FIFA and UEFA team ratings can be officially used to capture the level of each team and, indirectly, of each group (Groll et al., 2015, 2018b).

To conclude with, the vanilla model is not so bad in comparison with the feature-based performance models. It is simpler but requires an increased number of games from the current season in order to obtain reliable predictions. On the other hand it is not useful for the beginning of the season or for tournaments of other types. Also, data from previous seasons may be irrelevant and might not offer reliable estimates of the team abilities in the current season.

On the other hand, the performance-based models need additional number of information in the form of covariates but lower number of games. They can be used more efficiently at the beginning of the season and for tournaments of more complicated structure. Finally, data from previous seasons are also relevant and offer an increased number of precisions in the estimates.

2.1.5 Models for international and European club tournaments

For hybrid tournaments of mini-round-robin qualification rounds followed by elimination (knock-out) phases like the ones in tournaments of national teams (e.g. Euro or World cups) and international club competitions such as the UEFA Champions League or the Europa League, models with covariates are considerably better fitted than the vanilla models. Simple vanilla models are totally inappropriate when using data from the group phase in order to predict the outcomes in the knock-out phases. The reason is due to the fact that the attacking and defending parameters we estimate are relative to the strength of the other teams. So when the teams of different groups do not cross-over in games, the corresponding parameters are estimated only with regard of the opponents of each team. Hence, if a team is much better than its opponents in a specific group it will appear with exceptional attacking and defensive abilities due to the fact that the overall level of the group was lower than the corresponding performance level of the rest of the groups. In order to have a model which will work effectively, we need at least a common quantitative covariate which will reflect the quality of each team (usually these are UEFA or FIFA ratings, as in Groll et al. (2015, 2018b). We refer to Chapter 6 for further details about the modelling of these kinds of tournaments.

2.2 How to setup the data

2.2.1 Game-arranged data

As we have already mentioned at the beginning of this chapter, the natural arrangement of our dataset is to consider a $n \times (2p+2)$ data matrix, where n is the number of games under consideration and p is the number of covariates under consideration for each team. The goals scored by the two opponent teams will be denoted as Y_{i1} and Y_{i2}, where the former denotes the home team and the latter the away/guest team. From the covariates we need to consider, two are the most basic ones which are required to fit the vanilla model: the home team and the away (or guest) team. These variables will be

TABLE 2.1
Example of a game-arranged dataset

Game				Final Score		Team Covariates					
						Home			Away		
(i)	Opponent Teams	h_i	a_i	Y_{i1}	Y_{i2}	$X_{i1}^{(1)}$	$\dots$	$X_{ip}^{(1)}$	$X_{i1}^{(2)}$	$\dots$	$X_{ip}^{(2)}$
1	Man Und – Chelsea	10	2	1	1	$x_{11}^{(1)}$	$\dots$	$x_{1p}^{(1)}$	$x_{11}^{(2)}$	$\dots$	$x_{1p}^{(2)}$
2	Liverpool – Coventry	8	3	2	0	$x_{21}^{(1)}$	$\dots$	$x_{2p}^{(1)}$	$x_{21}^{(2)}$	$\dots$	$x_{2p}^{(2)}$
$\vdots$	$\vdots$	$\vdots$	$\vdots$	$\vdots$	$\vdots$	$\vdots$	$\vdots$	$\vdots$	$\vdots$	$\vdots$	$\vdots$
n	Teams of game n	h_n	a_n	Y_{n1}	Y_{n2}	$x_{n1}^{(1)}$	$\dots$	$x_{np}^{(1)}$	$x_{n1}^{(2)}$	$\dots$	$x_{np}^{(2)}$

considered as categorical with levels $k = 1, \dots, K$ and their level codes will be denoted as h_i and a_i; see Table 2.1 for an example of such dataset.

2.2.2 Univariate-arranged data

When we fit the standard double Poisson model, or generally models assuming conditional independence, then the data should be arranged in univariate fashion assuming one response: the number of goals scored by each team. Hence, the data of each game will be placed in two data rows. The final dataset will now have dimension of $2n$ rows and $p+3$ covariates (including the additional dummy of the home effect denoted by H). We will refer to this data arrangement as the univariate data arrangement. All data and related parameters under this setup will be denoted by using a "*" superscript to discriminate from the original, game-oriented approach, presented in Sections 2.1.1 and 2.1.2. It is also important to note that the team variables now refer to the scoring team and the team accepting the goals in contrast with the game-oriented approach where the covariates were referring to the home and away team. Finally, the home team is now indicated by using a dummy indicator variable called H. The parameters of these covariates will correspond to the attacking and defensive parameters (abilities) of the model and the corresponding covariates as attacking and defending teams. Hence for the first game of Table 2.1 we will consider the following values: Goals scored $Y_i = 1^*$, $H_i = 1$, Attacking/Scoring team A_i= 10 (Man Und), Defending Team D_i=2 (Chelsea) for the first row and the values $Y_i^* = 1$, $H_i = 0$, Attacking/Scoring team A_i= 2 (Chelsea) and Defending Team D_i= 10 (Man. Utd). Note that the attacking team and the defending team are reversed in the second row. Similarly, the covariates now will refer to the attacking and defending teams and in the second row they will be reversed; see last line of Table 2.2 for the general representation.

TABLE 2.2
Example of a univariate-arranged dataset

	Game	Scoring	Goals	Home	Attacking Team	Defending Team	Team Covariates					
							Attacking			Defending		
i	(G_i)	Team	(Y_i^*)	(H_i)	(A_i)	(D_i)	$X_{i1}^{*(1)}$	...	$X_{ip}^{*(1)}$	$X_{i1}^{*(2)}$	...	$X_{ip}^{*(2)}$
1	1	Man Und	1	1	10	2	$x_{11}^{(1)}$	...	$x_{1p}^{(1)}$	$x_{11}^{(2)}$	...	$x_{1p}^{(2)}$
2	1	Chelsea	1	0	2	10	$x_{11}^{(2)}$	...	$x_{1p}^{(2)}$	$x_{11}^{(1)}$	...	$x_{1p}^{(1)}$
3	2	Liverpool	2	1	8	3	$x_{21}^{(1)}$	...	$x_{2p}^{(1)}$	$x_{21}^{(2)}$	...	$x_{2p}^{(2)}$
4	2	Coventry	0	0	3	8	$x_{21}^{(2)}$	...	$x_{2p}^{(2)}$	$x_{21}^{(1)}$	...	$x_{2p}^{(1)}$
⋮	⋮	⋮	⋮	⋮	⋮	⋮	⋮	⋮		⋮	⋮	⋮
$2n-1$	n	Home Team n	Y_{n1}	1	HT_n	AT_n	$x_{n1}^{(1)}$	...	$x_{np}^{(1)}$	$x_{n1}^{(2)}$	...	$x_{np}^{(2)}$
$2n$	n	Away Team n	Y_{n2}	0	AT_n	HT_n	$x_{n1}^{(2)}$	...	$x_{np}^{(2)}$	$x_{n1}^{(1)}$	...	$x_{np}^{(1)}$

2.2.3 Model formulation for univariate-arranged data

Under the univariate-arranged data representation, the model formulation should be slightly modified. Hence the Poisson model is written as

$$\begin{aligned} Y_\imath^* &\sim \mathcal{P}oisson(\lambda_\imath^*) \\ \log \lambda_\imath^* &= \mu + \eta_\imath^{van} + \eta_\imath^{cov} \\ \eta_\imath^{van} &= home \times H_\imath + att_{A_\imath} + def_{D_\imath} \\ \eta_\imath^{cov} &= \sum_{i=1}^{p} \beta_j^{(1)} X_{\imath j}^{*(1)} + \sum_{i=1}^{p} \beta_j^{(2)} X_{\imath j}^{*(2)} \\ & \quad \text{for } \imath \in \{1, \dots, 2n\}. \end{aligned}$$

Note that $\imath = 2i+2-\ell$ with $i \in \{1, \dots, n\}$ and $\ell = 1, 2$. In the above formulation, $\eta_\imath^{van}$ is the part of the model referring to the vanilla representation, $\eta_\imath^{cov}$ is the linear predictor incorporating additional team covariates (usually based on performance). The simple vanilla model arises if $\eta_\imath^{cov}$ is removed from the above formulation, while usual predictive models based on performance arise using only this part in the model. Hybrid models combining both parts in the formulation are rarer in the literature since the abilities appearing in the vanilla component of the model are usually collinear with the performance features of the second part, hence they do not improve the predictive ability of the model when introduced additively.

In the above formulation, the model parameters can be denoted by the parameter vector

$$\boldsymbol{\theta} = \big(\mu, home, def_2, \dots, def_K, att_2, \dots, def_K, \beta_1^{(1)}, \dots, \beta_p^{(1)}, \beta_1^{(2)}, \dots, \beta_p^{(2)}\big).$$

2.3 Methods of estimation Part I: The classical approach and the maximum likelihood estimation

In modern statistical science, there are two dominant approaches for parameter estimation. In the following sections, we will shortly refer to these two main approaches used for estimation of the model parameters. The first is the classical approach based on the notion of likelihood function and the second is the Bayesian approach which is based on the notion of the posterior distribution. Since this book focuses on the practical side of the implementation of statistical models in football data, we will try to provide only the basic notions needed to understand the implementation of the methods. For more details we refer the reader to specialized books of Statistical inference. In this section will start by briefly describing the first approach.

2.3.1 The likelihood function

Traditionally, in statistical inference, model parameters are considered as fixed but unknown parameters which we wish to calculate for a given population. The way to do so is to consider a representative, randomly drawn part of the population, which is known as the sample. The data we usually analyze in any survey or study is a sample from a population. The aim is to (approximately) calculate the population parameters of the model from the sample. This is achieved by mathematical functions which are called "Estimators". Estimators are selected in a way that they have some good properties which ensure that when the sample is large then they will give you the correct parameter value. The value of the estimator from each sample is called estimate of the parameter. If we take multiple different samples, then we will have many different values of the estimator (i.e. different estimates). This induces that an Estimator is a random variable which takes different values for each sample. As a random variable it will be accompanied by a mean/expected value and a standard deviation which will reflect the variability of the estimator.

One of these desirable properties of an estimator is unbiasedness. An unbiased estimator has mean/expected value equal to the parameter of interest. So essentially it means that all estimates will be around the true value. Under this perspective, the standard deviation of an unbiased estimator captures how close the estimates will be in the true parameter value. Therefore it is the error of the estimator and (its inverse) captures the precision of the estimator. Hence, the standard deviation of an estimator is called its standard error. Therefore, we wish to have unbiased estimators with as small as possible standard errors. For this reason, when we compare different estimators we may select the one with the smallest standard error. Under this perspective, another desirable property of the estimators is consistency. An estimator will be consistent if it converges (in probability) to the parameter of interest as the sample size increases. In principle this means that our knowledge becomes more and more precise about the parameter of interest as the sample size increases.

So what is the likelihood function and why maximizing it is so important for statistical inference? In order to specify the likelihood, we firstly need to specify the notion of the statistical model. Let us assume a random response variable Y, in our context here the number of goals, the goal difference or the match outcome. For this match outcome variable we assume a probability distribution $f(y|\boldsymbol{\theta})$, i.e. a probabilistic rule describing the probabilities of all the possible outcomes. The vector $\boldsymbol{\theta} = (\theta_1, \ldots, \theta_p)$ captures the model parameters that we wish to estimate from the data. For example, in the vanilla Poisson model described in Sections 2.1.1 and 2.1.2, $\boldsymbol{\theta}$ will be including the constant term, the home effect and the defensive and defensive team abilities. The distribution $f(y|\boldsymbol{\theta})$ is also called the sampling distribution of the model and its shape or type is one of the main assumptions when we use parametric modelling to describe a stochastic phenomenon. Hence, a statistical model is mainly characterized by the sampling distribution we assume for the (response) variable of interest. For a set of random variables $\boldsymbol{Y} = (Y_1, \ldots, Y_n)$ representing the possible outcomes of a sample of n observations, the joint density or probability function for a given sample $\boldsymbol{y} = (y_1, \ldots, y_n)$ is given by

$$f(\boldsymbol{y}|\boldsymbol{\theta}) = f(y_1, \ldots, y_n|\boldsymbol{\theta}).$$

This joint distribution is further simplified to

$$f(\boldsymbol{y}|\boldsymbol{\theta}) = f(y_1|\boldsymbol{\theta}) \cdot f(y_2|\boldsymbol{\theta}) \cdot \ldots \cdot f(y_n|\boldsymbol{\theta}) = \prod_{i=1}^{n} f(y_i|\boldsymbol{\theta})$$

when the random variables Y_i are assumed to be independent and identically distributed (which is a realistic assumption for a number of popular models). In the following we will assume the case of independent and identically distributed random variables.

The likelihood function is nothing more that the sampling distribution when seen as a function of the parameters of interest for a given sample $\boldsymbol{y}$, that is

$$\mathcal{L}(\boldsymbol{\theta}) = f(y_1, \ldots, y_n|\boldsymbol{\theta}) = \prod_{i=1}^{n} f(y_i|\boldsymbol{\theta}).$$

Since the $\boldsymbol{y}$ is observed, we assume that its probability or its density $f(\boldsymbol{y}|\boldsymbol{\theta})$ will be high. So we work with the inverse logic and we consider as estimators $\widehat{\boldsymbol{\theta}}$ the functions we obtain if we maximize $f(\boldsymbol{y}|\boldsymbol{\theta})$, and therefore the likelihood function $\mathcal{L}$. It is proven that maximum likelihood estimators (MLEs) have a number of important asymptotic properties such as consistency, efficiency and normality. In principle this means that the MLEs estimate the true value for large samples with the smallest possible variance and moreover we can assume that its distribution is normal allowing for the implementation of a number of hypothesis tests. Moreover, we work in the log scale in all computations since, being the logarithm a continuous monotonic function, the maximum value will remain the same. Hence, in practice, we maximize the log-likelihood denoted by

$$\ell(\boldsymbol{\theta}) = \log f(y_1, \ldots, y_n|\boldsymbol{\theta}) = \sum_{i=1}^{n} \log f(y_i|\boldsymbol{\theta}).$$

2.3.2 Maximizing the likelihood

In order to obtain the maximum likelihood estimates (MLEs) $\widehat{\boldsymbol{\theta}}$, we need to maximize the likelihood by taking the first derivative of the log-likelihood and setting it equal to zero. Hence, the MLE $\widehat{\boldsymbol{\theta}}$ of $\boldsymbol{\theta}$ will be one of the values that solve the equation

$$\frac{d\ell(\boldsymbol{\theta})}{d\boldsymbol{\theta}} = 0$$

(an extra requirement is needed to decide if the implied point is a point of local maximum but we will not refer to it here to keep the text simple).

In order to understand how this can be used, we may consider as an example the simple normal linear regression model. Then the model is written as

$$Y_i \sim N(\alpha + \beta x_i, \sigma^2)$$

where $N(\mu, \sigma^2)$ is the normal distribution with mean μ and variance σ^2 and density function

$$f(y; \mu, \sigma^2) = \frac{1}{\sqrt{2\pi}\sigma} e^{-\frac{1}{2}\left(\frac{y-\mu}{\sigma}\right)^2}.$$

For this model, the parameters we wish to estimate are $\boldsymbol{\theta} = (\alpha, \beta, \sigma^2)$. Then, the log-likelihood will be given by

$$\begin{aligned} \ell(\boldsymbol{\theta}) &= \sum_{i=1}^{n} \log \left\{ \frac{1}{\sqrt{2\pi}\sigma} e^{-\frac{1}{2}\left(\frac{y_i-\alpha-\beta x_i}{\sigma}\right)^2} \right\} \\ &= -\frac{n}{2}\log(2\pi) - \frac{n}{2}\log(\sigma^2) - \frac{1}{2\sigma^2}\sum_{i=1}^{n}(y_i - \alpha - \beta x_i)^2. \end{aligned}$$

Taking the derivatives of the log-likelihood with respect of α, β and σ^2 we end up with the following MLEs

$$\begin{aligned} \widehat{\beta} &= \frac{\sum_{i=1}^{n}(x_i - \overline{x})(y_i - \overline{y})}{\sum_{i=1}^{n}(y_i - \overline{y})^2} \\ \widehat{\alpha} &= \overline{y} - \widehat{\beta}\overline{x} \\ \widehat{\sigma}^2 &= \frac{1}{n}\sum_{i=1}^{n}\sum_{i=1}^{n}(y_i - \widehat{\alpha} - \widehat{\beta}x_i)^2. \end{aligned}$$

The above is a relatively simple example where we can obtain the MLEs in closed form expressions using differential calculus. Nevertheless, in most cases we cannot have the MLEs in closed form expression. In such occasions, optimization algorithms are used. The most common methods in statistics are the Newton-Raphson algorithm and the Expectation-Maximization (EM) method which we will briefly describe in the following paragraphs.

The **Newton-Raphson algorithm** actually is not a method for maximization, but it is used to find the solution of an equation of the form $f(x) = 0$; thus, it can be used to solve the solution of $\ell'(\boldsymbol{\theta}) = 0$ finding by this way the local maxima of the likelihood method. The Newton-Raphson algorithm is the standard way to find the MLEs in generalized linear models. It is based on the simple iterative equation

$$x_t = x_{t-1} + \frac{f(x_{t-1})}{f'(x_{t-1})} \text{ for } t = 1, 2, \ldots$$

which will be stopped when $|x_t - x_{t-1}| < \epsilon$ or $|f(x_t)| < \epsilon'$ for some selected precision values ϵ and ϵ'. The implementation of Newton- in the univariate case is summarized as

---

Algorithm 1 Newton-Raphson Algorithm for the univariate case

Input: Set precision parameter $\epsilon' > 0$.

Initialize: Set initial value x_0 for x.

For $t = 1, 2, \ldots,$ REPEAT:

Set $x_t = x_{t-1} + \frac{f(x_{t-1})}{f'(x_{t-1})}$.

Stop for-loop if $|f(x_t)| < \epsilon'$.

End of loop-for.

Output: Report x_t.

For its implementation in statistics we will need to use the multivariate extension of this equation which involves both the first and the second derivatives of the log-likelihood with respect to each element of the parameter vector $\boldsymbol{\theta}$. Details are omitted for brevity; interested readers are prompt to the computational statistics book of Givens and Hoeting (2012, Section 2.2.1) and the online book of Peng (2022, Section 2.4.3). The uniroot function can be used to implement the method in R. Nevertheless, it can identify only one root inside a given interval. The uniroot.all function expands the uniroot function and can identify multiple roots, when they exist. This function is available in rootSolve package. A detailed description of the Newton-Raphson algorithm and of how you can use implement in R is available at `https://rpubs.com/aaronsc32/newton-raphson-method`.

Another maximization algorithm which is popular in a specific class of models in statistics is the Expectation-Maximization (EM) method. This method is appropriate when some auxiliary or latent data/variables are a key component of the model being estimated. In other situations, a model might be rewritten in this format just to simplify computations and enable the use of this computational algorithm (similar approaches are also employed in the implementation of Markov Chain Monte Carlo methods for Bayesian inference). The extra data or random variables in these cases are referred to as auxiliary because they are nonsense for the model interpretation and they are only introduced in the model to make computations easier. Within this framework, these auxiliary or latent variables/data are also called missing variables/data. EM can be implemented in factor analysis and latent variable models, in random effects and mixed models, in cluster analysis and mixture models.

The EM algorithm is an iterative algorithm. algorithm The method was originally introduced by Dempster et al. (1977) and it is popular due to its simplicity. The method comprises by two simple steps: E-step and the M-step. In the E-step, we estimate the latent/auxiliary data by calculating their expectations, whereas in the M-step we maximize the model parameters given that latent/auxiliary data are equal to their expected values found in the E-step.

In the general approach, the EM algorithm will be useful when the model sampling distribution $f(\boldsymbol{y}|\boldsymbol{\theta})$ is difficult to be handled computationally (or it is even unavailable in closed form) and the corresponding likelihood can be easily maximized. In order to work with the EM algorithm we need to identify auxiliary variables $\boldsymbol{Z}$ where we can re-write the model's sampling distribution as the marginal distribution when you ignore (or integrate out) $\boldsymbol{Z}$. Therefore, we need to be able to re-write $f(\boldsymbol{y}|\boldsymbol{\theta})$ in the following way

$$f(\boldsymbol{y}|\boldsymbol{\theta}) = \int f(\boldsymbol{y}, \boldsymbol{Z}|\boldsymbol{\theta})d\boldsymbol{Z} = \int f(\boldsymbol{y}|\boldsymbol{Z}, \boldsymbol{\theta})f(\boldsymbol{Z}|\boldsymbol{\theta})d\boldsymbol{Z}.$$

In the E-step, we calculate the expectation of the log-likelihood using the full data (observed and missing/auxiliary) given the observed ones. Hence, we calculate

$$\mu(\boldsymbol{\theta}) = \mathrm{E}\Big[\log f(\boldsymbol{y}|\boldsymbol{Z}, \boldsymbol{\theta})\Big|\boldsymbol{y}, \boldsymbol{\theta}\Big] = \int \Big[\log f(\boldsymbol{y}, \boldsymbol{Z}|\boldsymbol{\theta})\Big] f(\boldsymbol{Z}|\boldsymbol{y}, \boldsymbol{\theta})d\boldsymbol{Z},$$

and then we maximize $\mu(\boldsymbol{\theta})$ with respect to $\boldsymbol{\theta}$. The algorithm is repeated until $\boldsymbol{\theta}$ and $\mu(\boldsymbol{\theta})$ stabilize and there are minor changes between iterations.

Other methods also exist, especially in machine learning literature where the target function is extremely complicated; see for example reinforcement algorithms for machine learning methods.

2.4 Illustration: Fitting the double Poisson model with MLE approach

In this section we are going to use the data from the matches of the English Premier League 2006–2007 season. This dataset will serve as an example for modelling football match data using the double Poisson model (see Eq. 2.2 and 2.3) using the MLE approach. The analysis employs a simple vanilla model, which assumes a Poisson distribution for goals scored by each team. This dataset, accessible through the engsoccerdata package in R, provides match results and team performance data in a structured format. This approach highlights how basic statistical models can provide insights into team performance while remaining interpretable and straightforward to implement.

Before fitting the model, the dataset requires transformation (as described in Section 2.2.2) to meet the requirements of a Poisson generalized linear model (GLM). Each match is represented as two rows in the dataset, one for each team, with the response variable indicating the number of goals scored ("goals1" and "goals2"). This format ensures that the model accounts for both home and away performances, enabling the inclusion of predictors such as a home-team indicator, which captures the advantage typically associated with playing on home ground. This transformation not only aligns with the structure required for GLMs but also enhances the model's ability to assess critical factors influencing match outcomes, such as home advantage. Code-Snippet 1 provides the R code for preparation of the data, while R Outputs 1 and 2 display samples of the original dataset and the transformed dataset, respectively. The transformed dataset is now ready for use with the glm function in R.

Code Snippet 1 Data transformation for use with the glm function in R

```
n<-nrow(chap07_ex2_soccer)
all(levels(chap07_ex2_soccer$ht)==levels(chap07_ex2_soccer$at))
goals <-c(chap07_ex2_soccer$goals1,chap07_ex2_soccer$goals2)
game <- c(1:n, 1:n )
home <- c( rep(1,n), rep(0,n)  )
att <- factor(c(chap07_ex2_soccer$ht, chap07_ex2_soccer$at) )
def <- factor(c(chap07_ex2_soccer$at, chap07_ex2_soccer$ht))
levels(att) <- levels(chap07_ex2_soccer$ht)

premier <- data.frame( game=game, att=att, def=def, home=home, goals=goals)
head(premier)
i<-order(game)
premier<-premier[i,]
head(premier)
```

```
> head(chap07_ex2_soccer)
        team1 goals1 goals2          team2          ht            at z
1 Sheff Utd        1      1 Liverpool       Sheff Utd      Liverpool 0
2 Arsenal          1      1 Aston Villa       Arsenal    Aston Villa 0
3 Everton          2      1 Watford           Everton        Watford 1
4 Newcastle        2      1 Wigan           Newcastle          Wigan 1
5 Portsmouth       3      0 Blackburn      Portsmouth      Blackburn 3
6 Reading          3      2 Middlesbrough     Reading Middlesbrough 1
```

Output 1: Dataset before transformation (one row per match).

```
> head(premier)
    game         att         def home goals
1      1   Sheff Utd   Liverpool    1     1
381    1   Liverpool   Sheff Utd    0     1
2      2     Arsenal Aston Villa    1     1
382    2 Aston Villa     Arsenal    0     1
3      3     Everton     Watford    1     2
383    3     Watford     Everton    0     1
```

Output 2: Final dataset for use with the glm function in R (two rows per match).

Once the dataset is structured, the model is fitted in R using the glm function. This process involves specifying the Poisson family, which is suitable for count data like goals. The simplicity of the vanilla model allows for quick computation and easy interpretation of results, making it a practical choice for analyzing sports data. For instance, the coefficients of the model can help quantify the home advantage or identify teams with particularly strong offensive or defensive performances.

Code-Snippet 2 provides the details about fitting the double Poisson model using the glm function in R after imposing the sum-to-zero constraints on the team attacking and defensive parameters/abilities. The resulted model coefficients (constant, home effect, attacking and defensive abilities) are presented in R–Output 3.

Code Snippet 2 Fitting the double Poisson model using the glm function.

```
# Setting the sum-to-zero constraints for attacking parameters
contrasts(premier$att)<-contr.sum(20)
# Setting the sum-to-zero constraints for defensive parameters
contrasts(premier$def)<-contr.sum(20)
# Fitting the Double Poisson model
model <- glm( goals~home+att+def, family=poisson, data=premier )
# Summarizing the Double Poisson model
summary(model)
# Double Poisson model coefficients
round(summary(model)$coef,3)
```

```
> round(summary(model)$coef,3)
            Estimate Std. Error z value Pr(>|z|)
(Intercept)   -0.075      0.053  -1.413    0.158
home           0.376      0.067   5.637    0.000
att1           0.329      0.124   2.646    0.008
att2          -0.047      0.149  -0.314    0.753
att3           0.159      0.136   1.165    0.244
att4           0.055      0.143   0.385    0.701
att5          -0.261      0.166  -1.566    0.117
att6           0.333      0.123   2.693    0.007
att7           0.138      0.136   1.015    0.310
att8          -0.149      0.158  -0.945    0.345
att9           0.220      0.130   1.688    0.091
att10         -0.438      0.180  -2.440    0.015
att11          0.597      0.110   5.438    0.000
att12         -0.015      0.147  -0.100    0.921
att13         -0.164      0.158  -1.039    0.299
att14          0.000      0.146  -0.001    1.000
att15          0.151      0.136   1.106    0.269
att16         -0.327      0.171  -1.910    0.056
att17          0.251      0.130   1.923    0.055
att18         -0.421      0.180  -2.345    0.019
att19         -0.233      0.164  -1.418    0.156
def1          -0.231      0.164  -1.409    0.159
def2          -0.095      0.152  -0.627    0.531
def3           0.191      0.134   1.428    0.153
def4           0.147      0.136   1.083    0.279
def5           0.276      0.127   2.169    0.030
def6          -0.608      0.197  -3.092    0.002
def7          -0.216      0.162  -1.332    0.183
def8           0.281      0.127   2.204    0.027
def9          -0.498      0.186  -2.680    0.007
def10         -0.040      0.147  -0.274    0.784
def11         -0.469      0.186  -2.520    0.012
def12          0.084      0.140   0.603    0.546
def13          0.036      0.143   0.251    0.802
def14         -0.069      0.150  -0.459    0.646
def15          0.052      0.143   0.362    0.717
def16          0.187      0.133   1.408    0.159
def17          0.197      0.134   1.471    0.141
def18          0.254      0.128   1.977    0.048
def19          0.260      0.128   2.030    0.042
```

Output 3: Double Poisson estimated parameters using the glm function in R.

As you may notice, on both attacking and defensive coefficients the parameter corresponding to the 20th team is missing due to the imposed constraints. We may restructure these parameters and add the missing parameter using the R syntax of Code-Snippet 3 resulting in the more structured table presented in R–Output 4.

From the table of R–Output 4, we can see, for example, that Manchester United had the highest attacking ability equal to 0.597 which means that it is expected to score 81.7% number of goals than an average team playing against the same opponent[1]. Postmouth attacking ability is exactly equal to zero

[1]This is calculated by considering $100 \times (e^{0.597} - 1)$.

Code Snippet 3 Code for restructuring attacking and defensive parameters.

```
abilities <- matrix( nrow=20,ncol=4 )
abilities[1:19,1:2] <- summary(model)$coef[2+1:19,1:2]
abilities[1:19,2:3] <- summary(model)$coef[21+1:19,1:2]
# Calculation of the abilities for the 20th team
abilities[20,1] <- -sum(summary(model)$coef[2+1:19,1])
abilities[20,3] <- -sum(summary(model)$coef[21+1:19,1])

# Adding team names
rownames(abilities)<-levels(premier$att)
# Adding column description
colnames(abilities)<-c( "Att", "SD-Att", "Def", "SD-Def" )
abilities
```

```
> round(abilities,3)
                 Att SD-Att    Def SD-Def
Arsenal        0.329  0.124 -0.231  0.164
Aston Villa   -0.047  0.149 -0.095  0.152
Blackburn      0.159  0.136  0.191  0.134
Bolton         0.055  0.143  0.147  0.136
Charlton      -0.261  0.166  0.276  0.127
Chelsea        0.333  0.123 -0.608  0.197
Everton        0.138  0.136 -0.216  0.162
Fulham        -0.149  0.158  0.281  0.127
Liverpool      0.220  0.130 -0.498  0.186
Man City      -0.438  0.180 -0.040  0.147
Man Utd        0.597  0.110 -0.469  0.186
Middlesbrough -0.015  0.147  0.084  0.140
Newcastle     -0.164  0.158  0.036  0.143
Portsmouth     0.000  0.146 -0.069  0.150
Reading        0.151  0.136  0.052  0.143
Sheff Utd     -0.327  0.171  0.187  0.133
Tottenham      0.251  0.130  0.197  0.134
Watford       -0.421  0.180  0.254  0.128
West Ham      -0.233  0.164  0.260  0.128
Wigan         -0.177     NA  0.263     NA
```

Output 4: Attacking and defensive parameters.

which means that it is an average team in terms of scoring. Manchester city is the worst with coefficient equal to -0.438 which means that it is expected to score 35% lower number of goals than an average team against the same opponent. Wigan (the team whose abilities were omitted in the original glm output) has negative coefficient which means that its scoring ability is less than average (by 16%).

2.5 Methods of estimation Part II: A short introduction to model-based Bayesian inference

Bayesian theory adopts a different approach from classical statistical theory in dealing with unknown parameters. In Bayesian theory, any unknown parameter is treated as a random variable and, therefore, it requires a prior distribution that reflects the prior knowledge or beliefs about the parameter. Therefore, interest lies in calculating the posterior distribution of the unknown parameters which incorporate both prior $f(\boldsymbol{\theta})$ and data information which is expressed by the likelihood $f(\boldsymbol{y}|\boldsymbol{\theta})$. Then, using a modern version of the Bayes theorem, the posterior distribution of the model parameters $\boldsymbol{\theta}$ is given by

$$f(\boldsymbol{\theta}|\boldsymbol{y}) = \frac{f(\boldsymbol{y}|\boldsymbol{\theta})f(\boldsymbol{\theta})}{f(\boldsymbol{y})} = \frac{f(\boldsymbol{y}|\boldsymbol{\theta})f(\boldsymbol{\theta})}{\int f(\boldsymbol{y}|\boldsymbol{\theta})f(\boldsymbol{\theta})d\boldsymbol{\theta}} \propto f(\boldsymbol{y}|\boldsymbol{\theta})f(\boldsymbol{\theta}). \tag{2.5}$$

Equation (2.5) turns the prior information into posterior via the use of data and the model likelihood.

The posterior distribution can be summarized by its moments or other summary statistics such as the posterior mean, the median, the quantiles, or the standard deviation, and so on. These measures can be used for inference concerning the parameter of interest.

When no prior information is available, one can use a variety of non-informative vague priors; see for examples of non-informative priors in Kass and Wasserman (1995) and Yang and Berger (1996).

However, in many occasions, the posterior distribution may not have a closed-form expression. A common way to overcome this difficulty is to use conjugate prior distributions. Such priors have the property that the resulting posterior distribution belongs to the same family of the prior distribution. Bernardo and Smith (2000) provide a comprehensive overview of conjugate priors. Alternatively, one can use asymptotic approximations, such as the Laplace approximation (see, for example, in Erkanli, 1994; Tierney and Kadane, 1986; Tierney et al., 1989), or numerical integration techniques (see, for example, in Evans and Swartz, 1996).

With the change of the century and the rapid advancement of computing power, Markov Chain Monte Carlo (MCMC) techniques have become quite popular and now they are considered as a standard computational tool for medium sized statistical models. These techniques generate samples from the posterior distribution, allowing for the accurate estimation of the posterior densities and the implementation of complex models for real life problems. We briefly describe these methodologies in Section 2.5.1 which follows.

2.5.1 Markov Chain Monte Carlo methods

Markov chain Monte Carlo (MCMC) methods are powerful algorithms that emerged in the statistical science in the 1990s and facilitated the widespread implementation of Bayesian data analysis after 2000s. The main reason for the popularity of these algorithms is their ability to estimate (indirectly) high dimensional integrals involved in the Bayesian computation of statistical models describing common real life problems. This has enabled the application of sophisticated models in many practical domains.

MCMC methods were first introduced by Metropolis et al. (1953), but they became widely used in statistical science after the works of Gelfand and Smith (1990) and Gelfand et al. (1990); see in Gilks et al. (1996) for a comprehensive overview of MCMC methods and their application in a variety of applications. The WinBUGS software and its descendants (OpenBUGS, MultiBUGS, Jags, NIMBLE, Stan), originally developed by Professor Spiegelhalter and presented in Spiegelhalter et al. (1996), offer an easy-to-use platform for applying Bayesian models via MCMC.

A Markov chain is a stochastic process $\{\boldsymbol{\theta}^{(1)}, \boldsymbol{\theta}^{(2)}, \cdots, \boldsymbol{\theta}^{(t)}\}$ such that $f(\boldsymbol{\theta}^{(t+1)}|\boldsymbol{\theta}^{(t)}, \cdots, \boldsymbol{\theta}^{(1)}) = f(\boldsymbol{\theta}^{(t+1)}|\boldsymbol{\theta}^{(t)})$. That is, the distribution of $\boldsymbol{\theta}$ in time $t+1$ given all the preceding $\boldsymbol{\theta}$ (for times t, $t-1$, ..., 1) depends only on $\boldsymbol{\theta}^{(t)}$. Moreover, $f(\boldsymbol{\theta}^{(t+1)}|\boldsymbol{\theta}^{(t)})$ is independent of time t. When the Markov chain is irreducible, aperiodic and positive recurrent, as $t \to \infty$ the distribution of $\boldsymbol{\theta}^{(t)}$ tends to its equilibrium distribution which is independent of the initial $\boldsymbol{\theta}^{(0)}$; for details see Gilks et al. (1996).

In order to generate a sample from $f(\boldsymbol{\theta}|\boldsymbol{y})$ we must construct a Markov chain with two desired properties. First, $f(\boldsymbol{\theta}^{(t+1)}|\boldsymbol{\theta}^{(t)})$ should be "easy to generate from" and, second, the equilibrium distribution of the selected Markov chain should be our target posterior distribution $f(\boldsymbol{\theta}|\boldsymbol{y})$.

We construct a Markov chain with the above requirements, then we select an initial value $\boldsymbol{\theta}^{(0)}$ and generate values until the equilibrium distribution is reached. The next step is to cut off the first B observations as a burn-in period in order to eliminate any possible effect of the initial values $\boldsymbol{\theta}^{(0)}$. The final sample of T (generated) observations will be $\{\boldsymbol{\theta}^{(B+1)}, \boldsymbol{\theta}^{(B+2)}, \cdots, \boldsymbol{\theta}^{(B+T)}\}$. Convergence of the MCMC can be checked by various methods; for details see (Cowles and Carlin, 1996) and Brooks and Roberts (1998). CODA (Plummer et al., 2006) and BOA (Smith, 2007) R packages can be used to apply certain diagnostic tests in order to check the convergence of the MCMC algorithm for a given generated sample.

Two are the most popular MCMC methods: Metropolis Hastings (Metropolis et al., 1953; Hastings, 1970) and the Gibbs sampler (Geman and Geman, 1984a).

2.5.1.1 The Metropolis-Hastings algorithm

In Metropolis-Hastings algorithm we follow iteratively three steps:

1. Generate $\boldsymbol{\theta}'$ from a proposal distribution $q(\boldsymbol{\theta}|\boldsymbol{\theta}^{(t)})$.
2. Calculate

$$\alpha = \min\left(1, \frac{f(\boldsymbol{y}|\boldsymbol{\theta}')f(\boldsymbol{\theta}')q(\boldsymbol{\theta}^{(t)}|\boldsymbol{\theta}')}{f(\boldsymbol{y}|\boldsymbol{\theta})f(\boldsymbol{\theta})q(\boldsymbol{\theta}'|\boldsymbol{\theta}^{(t)})}\right).$$

3. Update $\boldsymbol{\theta}^{(t+1)} = \boldsymbol{\theta}'$ with probability α, otherwise set $\boldsymbol{\theta}^{(t+1)} = \boldsymbol{\theta}^{(t)}$.

A common choice for the proposal is to consider a normal distribution centred in the parameter value of the previous iteration, that is $q(\boldsymbol{\theta}'|\boldsymbol{\theta}^{(t)}) = f_N(\boldsymbol{\theta}'; \boldsymbol{\theta}, \Sigma_\theta)$; where $f_N(\boldsymbol{x}; \boldsymbol{\mu}, \Sigma_\theta)$ is the probability density function of the multivariate normal distribution with mean $\boldsymbol{\mu}$ and variance-covariance matrix Σ_θ evaluated at $\boldsymbol{x}$. This simplifies the acceptance probability of Step 2 in the following expression

$$\alpha = \min\left(1, \frac{f(\boldsymbol{y}|\boldsymbol{\theta}')f(\boldsymbol{\theta}')}{f(f(\boldsymbol{y}|\boldsymbol{\theta})f(\boldsymbol{\theta})}\right).$$

The covariance matrix Σ_θ is a tuning parameter of the MCMC algorithm and controls the convergence speed of the algorithm. algorithm

Another standard choice is the independence sampler where the proposal distribution does not depend on the current state $\boldsymbol{\theta}^{(t)}$ of the chain, while the most frequent implementation is the single component Metropolis-Hastings where only one parameter is updated in each iteration.

2.5.1.2 Gibbs sampler

Geman and Geman (1984a) introduced the Gibbs sampler. Using the matching conditional posterior, we update one component in each step of this algorithm. algorithm For a given state of the chain $\boldsymbol{\theta}^{(t)}$, the steps of the algorithm are as follows:

$$\begin{array}{lll}
\text{Generate } \theta_1^{(t+1)} & \text{from} & f(\theta_1|\theta_2^{(t)}, \theta_3^{(t)}, \cdots, \theta_p^{(t)}, \boldsymbol{y}),\\
\text{Generate } \theta_2^{(t+1)} & \text{from} & f(\theta_2|\theta_1^{(t+1)}, \theta_3^{(t)}, \cdots, \theta_p^{(t)}, \boldsymbol{y}),\\
\text{Generate } \theta_3^{(t+1)} & \text{from} & f(\theta_3|\theta_1^{(t+1)}, \theta_2^{(t+1)}, \cdots, \theta_p^{(t)}, \boldsymbol{y}),\\
\vdots & \vdots & \vdots\\
\text{Generate } \theta_p^{(t+1)} & \text{from} & f(\theta_p|\theta_1^{(t+1)}, \theta_2^{(t+1)}, \cdots, \theta_{p-1}^{(t+1)}, \boldsymbol{y}),
\end{array}$$

where p is the number of components of the parameter vector $\boldsymbol{\theta}$. The generation from $f(\theta_j|\boldsymbol{\theta}_{\setminus j}, \boldsymbol{y}) = f(\theta_j|\theta_1^{(t+1)}, \cdots, \theta_{j-1}^{(t+1)}, \theta_{j+1}^{(t)}, \cdots, \theta_p^{(t)}, \boldsymbol{y})$ is relatively easy since it is a univariate distribution and can be written as $f(\theta_j|\boldsymbol{\theta}_{\setminus j}, \boldsymbol{y}) \propto f(\boldsymbol{\theta}|\boldsymbol{y})$ where all the variables except θ_j are held constant at their given values.

If in the Metropolis-Hastings algorithm we consider the full conditional posterior distribution $f(\theta_j|\boldsymbol{\theta}_{\setminus j}, \boldsymbol{y})$ as the proposal density $q(\boldsymbol{\theta}'|\boldsymbol{\theta}^{(t)})$, then we always accept the proposed move (with probability equal to one). This makes the Gibbs sampler to be a special case of the single component Metropolis-Hastings algorithm. algorithm The Gibbs sampler was used in a variety of applications in several fields. The development of WinBUGS (Lunn et al., 2000) also helped towards this direction.

2.5.1.3 Hamiltonian Monte Carlo

Hamiltonian Monte Carlo (HMC) (Betancourt and Girolami, 2015a) is an MCMC technique for sampling from complex probability distributions using concepts from physics. Through the derivatives of the target posterior distribution, it produces effective transitions over the whole space of the posterior of interest; for more details we refer to Betancourt and Girolami (2015b). By utilizing the density function derivatives, HMC can suggest distant moves that are more likely to be accepted, reducing by this way the correlation between subsequent samples and increasing the efficiency of the sampling process. Being a specific type of an MCMC technique, HMC produces a series of random samples that converge to the target distribution. HMC is extremely useful when sampling from high-dimensional and multimodal distributions, such those found in Bayesian inference problems. In a nutshell, HMC uses an approximate Hamiltonian dynamics simulation scheme based on numerical integration which is then corrected by performing a Metropolis acceptance step.

The primary rationale behind using HMC instead of conventional MCMC it the fact that HMC has the advantage that it can escape from regions of local maxima more easily. HMC is generally more efficient than standard MCMC methods, as it produces samples with lower auto-correlation and converges faster to the target posterior distribution. HMC can provide more effective moves that can travel through the space more successfully by taking advantage of the structure and the geometry of the target distribution.

Nevertheless, HMC has certain limitations and disadvantages. The most important is that the target distribution needs to be differentiable. This makes the method inappropriate for sampling from posterior distributions of discrete parameters. Another drawback is that HMC is prone to instabilities, which can lead to the algorithm diverging or producing inaccurate samples. Therefore, HMC must be carefully implemented and monitored in order to obtain accurate samples from the posterior distribution of interest.

The General algorithm

Although, HMC is much more complicated than standard MCMC methods described in Sections 2.5.1.1 and 2.5.1.2, we will try to describe the main steps of the algorithm as simple as possible. The HMC is summarized by the following steps

- **Step 1**: Introduction of an auxiliary momentum variable and generation of it
- **Step 2**: Setting up the Hamiltonian, and solve the induced differential equation using the Leapfrog Integrator
- **Step 3**: Metropolis accept step

So let us assume that we wish to generate samples from the posterior distribution $f(\boldsymbol{\theta}|\boldsymbol{y})$. Then in Step 1, we introduce a set of auxiliary momentum variables $\boldsymbol{\rho}$ and we sample from the joint posterior

$$f(\boldsymbol{\rho}, \boldsymbol{\theta}|\boldsymbol{y}) = f(\boldsymbol{\rho}|\boldsymbol{\theta}, \boldsymbol{y}) f(\boldsymbol{\theta}|\boldsymbol{y}).$$

A simplified version is to assume that $\boldsymbol{\rho}$ and $\boldsymbol{\theta}$ are independent and use a multivariate normal distribution for the momentum variable.

Next, in **Step 2**, the joint density $f(\boldsymbol{\rho}, \boldsymbol{\theta}|\boldsymbol{y})$ defines a Hamiltonian by setting

$$H(\boldsymbol{\rho}, \boldsymbol{\theta}) = -\log f(\boldsymbol{\rho}, \boldsymbol{\theta}|\boldsymbol{y}) = -\log f(\boldsymbol{\rho}|\boldsymbol{\theta}, \boldsymbol{y}) - \log f(\boldsymbol{\theta}|\boldsymbol{y}).$$

The first term $T(\boldsymbol{\rho}|\boldsymbol{\theta}) = -\log f(\boldsymbol{\rho}|\boldsymbol{\theta}, \boldsymbol{y})$ is called the kinetic energy while the second one is the potential energy and it is often denoted by $V(\boldsymbol{\theta})$). So in this step we first generate $\boldsymbol{\rho}$ from $f(\boldsymbol{\rho}|\boldsymbol{\theta}, \boldsymbol{y})$ and then we update a set of values (trajectory) denoted by $\boldsymbol{\rho}_t$ and $\boldsymbol{\theta}_t$ (for $t = 0, 1, \ldots, n$) by the Hamiltonian's differential equations

$$\begin{aligned}
\frac{d\boldsymbol{\theta}_t}{dt} &= -\frac{1}{f(\boldsymbol{\rho}_t|\boldsymbol{\theta}_t, \boldsymbol{y})} \frac{\partial f(\boldsymbol{\rho}_t|\boldsymbol{\theta}_t, \boldsymbol{y})}{\partial \boldsymbol{\rho}_t}, \\
\frac{d\boldsymbol{\rho}_t}{dt} &= +\frac{1}{f(\boldsymbol{\rho}_t|\boldsymbol{\theta}_t, \boldsymbol{y})} \frac{\partial f(\boldsymbol{\rho}_t|\boldsymbol{\theta}_t, \boldsymbol{y})}{\partial \boldsymbol{\theta}_t} + \frac{1}{f(\boldsymbol{\theta}_t|\boldsymbol{y})} \frac{\partial f(\boldsymbol{\theta}_t|\boldsymbol{y})}{\partial \boldsymbol{\theta}_t}.
\end{aligned} \tag{2.6}$$

The above system of equations describes a deterministic motion or trajectory. In this trajectory, we start at point $\boldsymbol{\rho}_0$ and it is defined for any $t \geq 0$.

The differential equation system (2.6) is then solved using the leapfrog integrator which is a numerical integration algorithm. The algorithm requires to specify the of leap frog steps L and the step size $\varepsilon > 0$ being a small positive number.

Suppose the chain at the current MCMC step/iteration t is at state $(\boldsymbol{\theta}^{(t)}, \boldsymbol{\rho}^{(t)})$. We consider a sequence of values $(\boldsymbol{\theta}_{\ell\varepsilon}, \boldsymbol{\rho}_{\ell\varepsilon})$ for $\ell = 0, 1, \ldots, L$ with $\boldsymbol{\theta}_0 = \boldsymbol{\theta}^{(t)}$ and $\boldsymbol{\rho}_0 = \boldsymbol{\rho}^{(t)}$. For any given $(\boldsymbol{\theta}_{\ell\varepsilon}, \boldsymbol{\rho}_{\ell\varepsilon})$, the next values $(\boldsymbol{\theta}_{\ell\varepsilon+\varepsilon}, \boldsymbol{\rho}_{\ell\varepsilon+\varepsilon})$ in the sequence, are given by

$$\begin{aligned}
\boldsymbol{\rho}_{\ell\varepsilon+\frac{\varepsilon}{2}} &= \boldsymbol{\rho}_{\ell\varepsilon} + \frac{\varepsilon}{2} \frac{\partial \log f(\boldsymbol{\theta} = \boldsymbol{\theta}_{\ell\varepsilon}|\boldsymbol{y})}{\partial \boldsymbol{\theta}}, \\
\boldsymbol{\theta}_{\ell\varepsilon+\varepsilon} &= \boldsymbol{\theta}_{\ell\varepsilon} - \varepsilon \frac{\partial \log f(\boldsymbol{\rho} = \boldsymbol{\rho}_{\ell\varepsilon+\frac{\varepsilon}{2}}|\boldsymbol{\theta} = \boldsymbol{\theta}_{\ell\varepsilon}, \boldsymbol{y})}{\partial \boldsymbol{\rho}}, \\
\boldsymbol{\rho}_{\ell\varepsilon+\varepsilon} &= \boldsymbol{\rho}_{\ell\varepsilon+\frac{\varepsilon}{2}} + \frac{\varepsilon}{2} \frac{\partial \log f(\boldsymbol{\theta} = \boldsymbol{\theta}_{\ell\varepsilon+\varepsilon}|\boldsymbol{y})}{\partial \boldsymbol{\theta}}.
\end{aligned}$$

At the end we set $(\boldsymbol{\rho}^*, \boldsymbol{\theta}^*) = (\boldsymbol{\rho}_{L\varepsilon}, \boldsymbol{\theta}_{L\varepsilon})$. These values, obtained at the end of the leapfrog integrator, are, then, used as a proposal value in the Metropolis of Step 3. Thus, in Step 3, we accept the proposed move $(\boldsymbol{\rho}^*, \boldsymbol{\theta}^*)$ with probability

$$\alpha = \min\left(1, e^{-H(\boldsymbol{\rho}^*, \boldsymbol{\theta}^*) + H(\boldsymbol{\rho}, \boldsymbol{\theta})}\right) = \min\left(1, \frac{f(\boldsymbol{\rho}^*|\boldsymbol{\theta}^*, \boldsymbol{y}) f(\boldsymbol{\theta}^*|\boldsymbol{y})}{f(\boldsymbol{\rho}|\boldsymbol{\theta}, \boldsymbol{y}) f(\boldsymbol{\theta}|\boldsymbol{y})}\right).$$

A Simplified Version of HMC

A simplified and popular version of HMC is to assume that $\boldsymbol{\rho}$ and $\boldsymbol{\theta}$ are independent and use a multivariate normal distribution for the momentum variable. Hence, we set $f(\boldsymbol{\rho}|\boldsymbol{\theta}, \boldsymbol{y}) \propto \exp\left(-\frac{1}{2}\boldsymbol{\rho}^T \boldsymbol{\Sigma}_\rho^{-1} \boldsymbol{\rho}\right)$ and, in Step 1 of HMC algorithm, we generate $\boldsymbol{\rho} \sim N_{d_\rho}(\mathbf{0}, \boldsymbol{\Sigma}_\rho)$, where d_ρ is the dimension of $\boldsymbol{\rho}$. The variance-covariance matrix can be the identity matrix or can be estimated from a pilot run or from the burn-in period of HMC. Hence, the algorithm for this special case of HMC can be summarized by the following simplified algorithm

The algorithm is summarized as follows.

- **Step 1**: Generate $\boldsymbol{\rho}$ from $N_{d_\rho}(\mathbf{0}, \boldsymbol{\Sigma}_\rho)$.
- **Step 2**: Generate a set of proposed values $(\boldsymbol{\rho}^*, \boldsymbol{\theta}^*) = (\boldsymbol{\rho}_{L\varepsilon}, \boldsymbol{\theta}_{L\varepsilon})$ by creating a sequence of values $(\boldsymbol{\theta}_{\ell\varepsilon}, \boldsymbol{\rho}_{\ell\varepsilon})$ for $\ell = 0, 1, \ldots, L$ with $\boldsymbol{\theta}_0 = \boldsymbol{\theta}^{(t)}$ and $\boldsymbol{\rho}_0 = \boldsymbol{\rho}^{(t)}$. For any given $(\boldsymbol{\theta}_{\ell\varepsilon}, \boldsymbol{\rho}_{\ell\varepsilon})$, the next values $(\boldsymbol{\theta}_{\ell\varepsilon+\varepsilon}, \boldsymbol{\rho}_{\ell\varepsilon+\varepsilon})$ in the sequence, are given by
$$\begin{aligned} \boldsymbol{\rho}_{\ell\varepsilon+\frac{\varepsilon}{2}} &= \boldsymbol{\rho}_{\ell\varepsilon} + \frac{\varepsilon}{2}\frac{\partial \log f(\boldsymbol{\theta} = \boldsymbol{\theta}_{\ell\varepsilon}|\boldsymbol{y})}{\partial \boldsymbol{\theta}}, \\ \boldsymbol{\theta}_{\ell\varepsilon+\varepsilon} &= \boldsymbol{\theta}_{\ell\varepsilon} - \varepsilon \boldsymbol{\rho}_{\ell\varepsilon+\frac{\varepsilon}{2}} \\ \boldsymbol{\rho}_{\ell\varepsilon+\varepsilon} &= \boldsymbol{\rho}_{\ell\varepsilon+\frac{\varepsilon}{2}} + \frac{\varepsilon}{2}\frac{\partial \log f(\boldsymbol{\theta} = \boldsymbol{\theta}_{\ell\varepsilon+\varepsilon}|\boldsymbol{y})}{\partial \boldsymbol{\theta}}. \end{aligned}$$
- **Step 3** Implement a Metropolis-Hastings step using $(\boldsymbol{\rho}^*, \boldsymbol{\theta}^*)$ as the new proposed values. Thus, we accept the proposed move $(\boldsymbol{\rho}^*, \boldsymbol{\theta}^*)$ with probability
$$\alpha = \min\left(1, \frac{f(\boldsymbol{\theta}^*|\boldsymbol{y})}{f(\boldsymbol{\theta}|\boldsymbol{y})} \times \times \exp\Big(-\frac{1}{2}\boldsymbol{\rho}^{*T}\boldsymbol{\Sigma}_\rho^{-1}\boldsymbol{\rho}^* + \frac{1}{2}\boldsymbol{\rho}^T\boldsymbol{\Sigma}_\rho^{-1}\boldsymbol{\rho}\Big)\right).$$

2.5.2 Tools for fitting Bayesian models

WinBUGS/OpenBUGS and its descendants

WinBUGS is standalone software which implements MCMC methods for Bayesian inference. It is the descendant of BUGS (Bayesian inference Using Gibbs Sampling) project which started in 1989 by Professor David Spiegelhalter and his team. The whole project was originally based in MRC Biostatistics Unit, Cambridge, and, later, was also supported by Imperial College School of Medicine at St Mary's in London[2]. Although the original BUGS software was running in Linux and DOS operating system, WinBUGS was the first implementation of BUGS in windows OS and its first version was available on the web on 1997 (Lunn et al., 2000). WinBUGS has more than 30,000 downloads. The program can be run via R using the package R2WinBUGS. The maintenance and support of WinBUGS was officially terminated on February 2019 and the project was continued by its descendants: OpenBUGS, MultiBUGS, JAGS and Nimble[3].

Although nowadays Stan is considered the standard tool for implementing Bayesian inference, the BUGS community is still strong with many users finding easier to use the more "old fashioned" WinBUGS like software.

[2]https://www.mrc-bsu.cam.ac.uk/software/bugs/

[3]Links: https://www.mrc-bsu.cam.ac.uk/software/bugs/openbugs/, https://www.multibugs.org/, https://mcmc-jags.sourceforge.io/ and https://r-nimble.org/

Stan Software

The implementation of HMC via home-made code is challenging due to its complexity and its various tuning parameters that need to be specified by the user. A ready-to-use solution is offered by STAN software (Carpenter et al., 2017), which is a probabilistic programming language similar to WinBUGS and OpenBUGS which simulates values from the posterior distribution using HMC. HMC can be easily run from R or Python. Additionally, STAN offers optimization methods for (penalized) maximum likelihood problems, and variational Bayes techniques for approximate Bayesian inference for high-dimensional problems.

MCMCpack package in R

R offers a great variety of packages than can implement Bayesian methods; CRAN task view on Bayesian inference maintained by Park et al. (2023) report more than 200 packages with 10 of them to be of general implementation. Among them, the most popular one is the MCMCpack (Martin et al., 2011), which offers a big variety of ready-to-use models and general purposes functions where the user can specify his model formulation. Although, it gained attention at the beginning having a popular book to support it (Albert, 2009), MCMCpack never manages to reach the popularity of WinBUGS.

pyMC package in Python

PyMC is the most popular MCMC package in Python. It further offers the implementation variational Bayes algorithm. Since the authors of this book are not Python fans, they cannot express a direct opinion about PyMC as an alternative to the above R based tools.

2.6 Illustration (continued): Fitting the double Poisson model with the Bayesian approach

Following the illustration in Section 2.4, we use the English Premier League dataset of season 2006–2007 to fit the double Poisson model (see Eqs. 2.2 and 2.3) using the Bayesian approach. We implement the model in WinBUGS/OpenBUGS, though similar methods apply to other software, such as STAN. For a more thorough Bayesian implementation of this model, readers may refer to (Ntzoufras, 2009, Section 7.4.3, pp. 249–257), which provides an in-depth discussion of the methodology and its application in WinBUGS.

Note that when fitting the model in WinBUGS/OpenBUGS, there is no need to transform the data into a univariate Poisson model, as described in Section 2.2.2. Instead, the model can be specified directly using the approach

```
> head(chap07_ex2_soccer)
            team1 goals1 goals2          team2          ht            at z
1 Sheff Utd         1      1 Liverpool        Sheff Utd     Liverpool 0
2 Arsenal           1      1 Aston Villa        Arsenal   Aston Villa 0
3 Everton           2      1 Watford            Everton       Watford 1
4 Newcastle         2      1 Wigan            Newcastle         Wigan 1
5 Portsmouth        3      0 Blackburn       Portsmouth     Blackburn 3
6 Reading           3      2 Middlesbrough      Reading Middlesbrough 1
> premier_bayes<-chap07_ex2_soccer
> premier_bayes$ht <- as.numeric(premier_bayes$team1)
> premier_bayes$at <- as.numeric(premier_bayes$team2)
> head(premier_bayes)
            team1 goals1 goals2          team2 ht at z
1 Sheff Utd         1      1 Liverpool        16  9 0
2 Arsenal           1      1 Aston Villa       1  2 0
3 Everton           2      1 Watford           7 18 1
4 Newcastle         2      1 Wigan            13 20 1
5 Portsmouth        3      0 Blackburn        14  3 3
6 Reading           3      2 Middlesbrough    15 12 1
> premier_bayes<-premier_bayes[,c(1,4,5,6,2,3)]
> head(premier_bayes)
          team1          team2 ht at goals1 goals2
1 Sheff Utd      Liverpool     16  9      1      1
2 Arsenal        Aston Villa    1  2      1      1
3 Everton        Watford        7 18      2      1
4 Newcastle      Wigan         13 20      2      1
5 Portsmouth     Blackburn     14  3      3      0
6 Reading        Middlesbrough 15 12      3      2
> premier_winbugs<-premier_bayes[,3:6]
> head(premier_winbugs)
  ht at goals1 goals2
1 16  9      1      1
2  1  2      1      1
3  7 18      2      1
4 13 20      2      1
5 14  3      3      0
6 15 12      3      2
```

Output 5: Dataset transformation of the original dataset of Output 1 to the one (premier_winbugs) used for WinBUGS/OpenBUGS model.

in Eqs. 2.2 and 2.3. Thus, each game can be represented in a single row with four columns: the home team (HT_i) and away team (AT_i), each identified by a code number, and the final score (home and away goals, denoted as `goals1` and `goals2`). The data are presented in R–Output 5.

Under this representation, the model can be re-expressed as

$$\begin{aligned}
\text{Goals}_{ik} &\sim Poisson(\lambda_{ik}) && \text{for } k = 1, 2 \\
\log(\lambda_{i1}) &= \mu + \text{home} + a_{\text{HT}_i} + d_{\text{AT}_i} && \\
\log(\lambda_{i2}) &= \mu \qquad\qquad + a_{\text{AT}_i} + d_{\text{HT}_i} && \text{for } i = 1, 2, \ldots, n,
\end{aligned}$$

where n is the number of games, μ is a constant parameter; home is the home effect; HT_i and AT_i are the home and away teams, respectively, competing in the ith game; a_k and d_k are the attacking and defensive effects–abilities of k team for $k = 1, 2, \ldots, K$; and K is the number of teams in the dataset under consideration (here $K = 20$). Hence, the above likelihood part can be specified in WinBUGS using the following code

```
for (i in 1:n){
 # stochastic component
 goals1[i] ~ dpois( lambda1[i] )
 goals2[i] ~ dpois( lambda2[i] )
 # linear predictor
 log(lambda1[i]) <- mu + home + a[ ht[i] ] + d[ at[i] ]
 log(lambda2[i]) <- mu        + a[ at[i] ] + d[ ht[i] ]
}
```

For the attacking and defensive parameters (a_k and d_k), we impose sum-to-zero constraints to ensure model identifiability and facilitate comparisons of each team's abilities relative to an overall attacking and defensive level. Specifically, we set

$$\sum_{k=1}^{K} a_k = 0 \text{ and } \sum_{k=1}^{K} d_k = 0. \tag{2.7}$$

Under this parametrization, all parameters have clear interpretation. The parameter μ represents the overall log-expected goals scored in away games, while *home* quantifies the home advantage as the difference in log-expected goals when two equally strong teams compete. The attacking and defensive parameters, a_k and d_k, measure deviations from the league's average attacking and defensive strength levels.

A positive attacking parameter ($a_k > 0$) indicates that the team performs better offensively than an average team in the league, while a negative defensive parameter ($d_k < 0$) indicates stronger defensive performance compared to an average league team.

To impose (2.7) in WinBUGS, we set the ability parameters for one team (e.g., $k = 1$) as:

$$a_1 = -\sum_{k=2}^{K} a_k \text{ and } d_1 = -\sum_{k=2}^{K} d_k.$$

This is implemented in WinBUGS syntax as:

```
a[1] <- -sum( a[2:K] )
d[1] <- -sum( d[2:K] )
```

Prior distributions must be defined for the remaining parameters: μ, *home* and a_k, d_k for $k = 2, \ldots, K$. The usual normal low-information prior with zero mean and large prior variance are used here (with prior precision equal to 10^{-4}). This is specified in WinBUGS using the following syntax

```
# prior distributions
mu~dnorm(0,0.001)
home~dnorm(0,0.001)
for (i in 2:K){
 a[i]~dnorm(0,0.01)
 d[i]~dnorm(0,0.01)
}
```

The full WinBUGS code is given in Figure 2.1 while the final data structure for direct use from WinBUGS and the initial values are given in Figure 2.2

Note, that we can run this model directly with the R environment using either R2WinBUGS or the R2OpenBUGS R packages.

2.6.1 Results

Posterior summaries of the Double Poisson model parameters are presented in Table 2.3. Figures 2.3 and 2.4 illustrate the posterior credible intervals for the attacking and defensive parameters of each team. According to the estimated model parameters, Manchester United had the highest attacking strength, while Chelsea had the lowest (i.e., strongest) defensive parameter. For a match between two average teams, the expected number of goals is 1.32 for the home team and 0.90 for the away team, implying a 46% increase in scoring rate when playing at home.

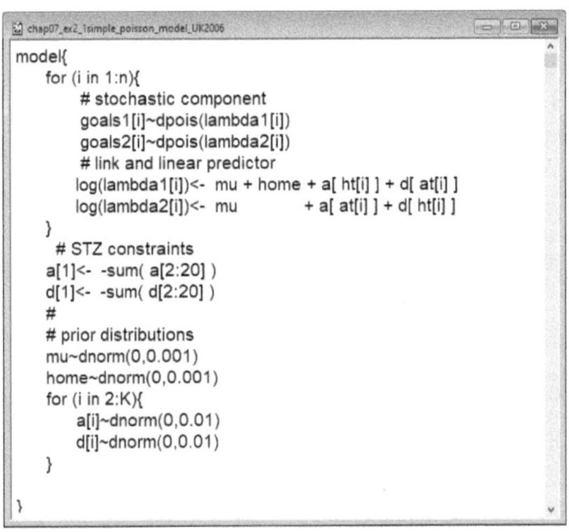

FIGURE 2.1
WinBUGS Syntax for the full double Poisson model.

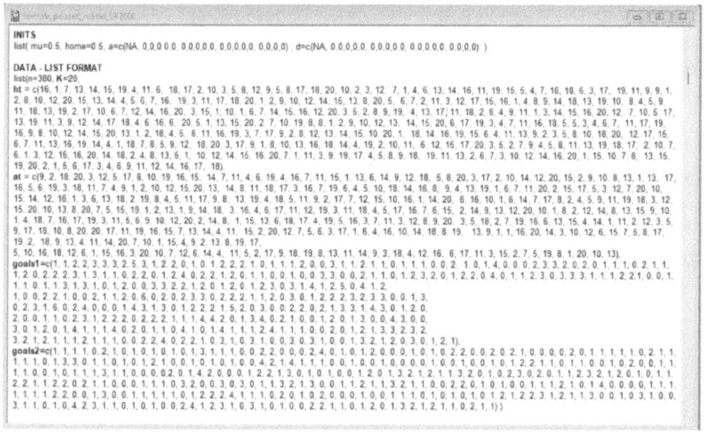

FIGURE 2.2
Initial values and data specification in WinBUGS/OpenBUGS model.

TABLE 2.3
Posterior summaries of model parameters for the Double Poisson model

		Posterior					Posterior			
				percentiles					percentiles	
Team[a]	Node	Mean	SD	2.5%	97.5%	Node	Mean	SD	2.5%	97.5%
1. Arsenal	a[1]	0.33	0.12	0.08	0.57	d[1]	–0.23	0.16	–0.57	0.08
2. Aston Villa	a[2]	–0.05	0.15	–0.34	0.23	d[2]	–0.10	0.16	–0.42	0.19
3. Blackburn	a[3]	0.16	0.14	–0.13	0.42	d[3]	0.20	0.14	–0.08	0.45
4. Bolton	a[4]	0.06	0.14	–0.23	0.32	d[4]	0.15	0.14	–0.13	0.41
5. Charlton	a[5]	–0.26	0.16	–0.61	0.04	d[5]	0.28	0.13	0.03	0.52
6. Chelsea	a[6]	0.33	0.12	0.08	0.56	d[6]	–0.62	0.20	–1.04	–0.25
7. Everton	a[7]	0.14	0.14	–0.13	0.41	d[7]	–0.22	0.17	–0.56	0.09
8. Fulham	a[8]	–0.14	0.16	–0.47	0.16	d[8]	0.29	0.13	0.02	0.53
9. Liverpool	a[9]	0.22	0.13	–0.04	0.47	d[9]	–0.51	0.19	–0.90	–0.16
10. Man City	a[10]	–0.44	0.18	–0.82	–0.09	d[10]	–0.04	0.15	–0.34	0.24
11. Man Utd	a[11]	0.60	0.11	0.38	0.81	d[11]	–0.47	0.19	–0.85	–0.12
12. Middlesbrough	a[12]	–0.02	0.15	–0.33	0.26	d[12]	0.09	0.14	–0.20	0.35
13. Newcastle	a[13]	–0.16	0.16	–0.48	0.13	d[13]	0.03	0.14	–0.26	0.31
14. Portsmouth	a[14]	0.00	0.15	–0.29	0.29	d[14]	–0.07	0.15	–0.37	0.22
15. Reading	a[15]	0.15	0.14	–0.13	0.40	d[15]	0.05	0.15	–0.24	0.33
16. Sheff Utd	a[16]	–0.33	0.17	–0.68	–0.01	d[16]	0.19	0.13	–0.08	0.44
17. Tottenham	a[17]	0.26	0.13	–0.00	0.51	d[17]	0.20	0.13	–0.06	0.45
18. Watford	a[18]	–0.42	0.18	–0.78	–0.07	d[18]	0.25	0.13	–0.01	0.50
19. West Ham	a[19]	–0.24	0.16	–0.56	0.07	d[19]	0.27	0.13	0.01	0.51
20. Wigan	a[20]	–0.19	0.16	–0.51	0.11	d[20]	0.27	0.13	0.01	0.51
	home	0.38	0.07	0.25	0.51	μ	–0.10	0.05	–0.20	0.003

Abbreviations: Man, Manchester; Utd, United; Sheff, Sheffield; Ham, Hampshire.

Posterior summaries of the predicted scores are presented in Table 2.4. For

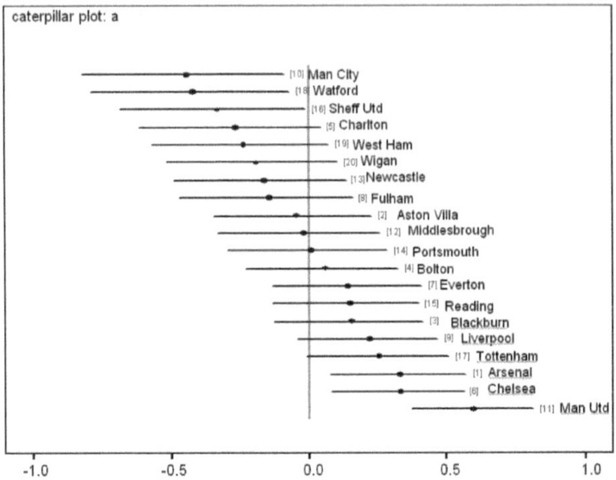

FIGURE 2.3
95% Posterior error bars of the attacking parameters.

both games, the posterior means suggest that the estimated model parameters are in agreement with the observed scores.

2.6.2 Prediction of future games.

In this section, we briefly demonstrate how to obtain predictions for two upcoming games. This approach can be easily extended to additional games using the same approach.

To implement this in WinBUGS, we replace the actual goals scored in the last two games (Tottenham vs. Manchester City and Watford vs. Newcastle) with `NA`; see Output 6. WinBUGS then automatically generates values for the

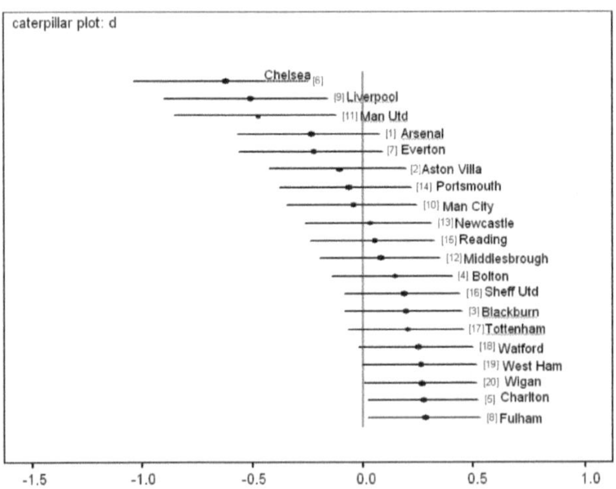

FIGURE 2.4
95% Posterior error bars of the defensive parameters.

TABLE 2.4
Posterior summaries of the expected scores for last two games of premier_winbugs dataset of Output 5

		Actual	Posterior		Posterior summaries of goal difference		
Home team	Away team	score	Median	Mean	Mean	SD	95% CI
379. Tottenham	Manchester City	2–1	1–1	1.65 – 0.72	0.93	1.59	(–2, 4)
380. Watford	Newcastle	1–1	1–1	0.93 – 1.02	-0.09	1.43	(–3, 3)

missing goals based on the predictive distribution and provides estimates for each score by monitoring the response variables `goals1` and `goals2`.

Another key quantity of interest is the probability of each match outcome (win/draw/loss), which can be easily handled in WinBUGS using the syntax in Code Snippet 4. In this syntax, the elements of `outcome` are binary indicators representing, respectively, a win, draw, or loss for the home team in each column.

Using similar syntax (see Code Snippet 5) we can also estimate the probabilities of the expected differences. In this syntax `pred.diff.counts` is a vector of binary random values indicating which difference appears in each

```
> tail(premier_bayes)
            team1           team2 ht at goals1 goals2
375 Man Utd       West Ham       11 19      0      1
376 Middlesbrough Fulham         12  8      3      1
377 Portsmouth    Arsenal        14  1      0      0
378 Sheff Utd     Wigan          16 20      1      2
379 Tottenham     Man City       17 10      2      1
380 Watford       Newcastle      18 13      1      1
> premier_bayes_missing <- premier_bayes
> premier_bayes_missing$goals1[379:380] <- NA
> premier_bayes_missing$goals2[379:380] <- NA
> tail(premier_bayes_missing)
            team1           team2 ht at goals1 goals2
375 Man Utd       West Ham       11 19      0      1
376 Middlesbrough Fulham         12  8      3      1
377 Portsmouth    Arsenal        14  1      0      0
378 Sheff Utd     Wigan          16 20      1      2
379 Tottenham     Man City       17 10     NA     NA
380 Watford       Newcastle      18 13     NA     NA
> premier_winbugs_missing<-premier_bayes_missing[,3:6]
> tail(premier_winbugs_missing)
    ht at goals1 goals2
375 11 19      0      1
376 12  8      3      1
377 14  1      0      0
378 16 20      1      2
379 17 10     NA     NA
380 18 13     NA     NA
```

Output 6: Introducing missing scores in the last two games of the dataset.

Code Snippet 4 WinBUGS syntax for the calculation outcome (win/draw/loss) probabilities of a game

```
# calculation of the predicted differences
pred.diff[1] <- goals1[379]-goals2[379]
pred.diff[2] <- goals1[380]-goals2[380]
#
# probability of each game outcome (win/draw/loss)
for (i in 1:2){
 outcome[i,1] <- 1 - step( -pred.diff[i] )   #home wins (diff>0)
 outcome[i,2] <- equals( pred.diff[i] , 0.0 )#draw (diff=0)
 outcome[i,3] <- 1-step( pred.diff[i] )      #home loses (diff<0)
}
```

MCMC iteration. Elements 2–12 denote differences from –5 to 5, while the first and the last elements of the vector denote differences lower than –5 and higher than 5, respectively.

Code Snippet 5 WinBUGS syntax for the calculation of the goal difference of a game

```
# calculation of the probability of each difference
for (i in 1:2){
 pred.diff.counts[i,1]<- 1-step(pred.diff[i]+5) # less than -5
 # equal to k-7 (-5 to 5)
 for (k in 2:12){
  pred.diff.counts[i,k]<-equals(pred.diff[i],k-7)}
 pred.diff.counts[i,13]<-step(pred.diff[i]-6) # greater than 5
}
```

Posterior probabilities of each predicted outcome and each value of the goal difference are summarized in Tables 2.5 and 2.6. Outcome probabilities indicate that Tottenham's probability of winning the game against Manchester City was about 60%, with a posterior mode of one goal difference. Concerning the second game (Watford vs. Newcastle), the posterior model probabilities confirm that the two teams have about equal probabilities of winning the game.

2.7 Tools for fitting football models in R

Over the last years, many R packages have been developed for fitting football models. One of the first attempts to provide ready-to-use code for fitting the bivariate Poisson model of Karlis and Ntzoufras (2003) was the bivpois package which is now available only in simple raw R code at `http://`

TABLE 2.5
Posterior probabilities of each game outcome for last two games of premier_winbugs dataset of Output 5

	Home team	Away team	Actual score	Posterior Probability Home wins	Draw	Away wins
379.	Tottenham	Manchester City	2–1	0.59	0.24	0.17
380.	Watford	Newcastle	1–1	0.33	0.30	0.37

TABLE 2.6
Posterior probabilities of each game goal difference for the last two games of premier_winbugs dataset of Output 5

	Home Team	Away Team	Actual Score	Posterior Probability of goal difference[a] ≤ -3	−2	−1	0	1	2	3	4	≥ 5
379.	Tottenham	Man City	2–1	0.012	0.036	0.119	0.242	**0.257**	0.185	0.090	0.037	0.023
380.	Watford	Newcastle	1–1	0.042	0.108	0.218	**0.303**	0.210	0.086	0.024	0.006	0.002

[a]Boldface indicates the maximum probability and the corresponding posterior mode of the difference.
Abbreviations: Man, Manchester.

`www2.stat-athens.aueb.gr/~jbn/papers/paper14.htm`. Also the code for the Poisson difference model (Skellam) of Karlis and Ntzoufras (2009) can be found at `http://www2.stat-athens.aueb.gr/~jbn/papers/paper20.htm`. Finally, Ntzoufras (2009) has fitted the double Poisson model on Italian Serie A data (see Example 7.2). Moreover, details (including code) for various other related models (such as the negative binomial, the generalized Poisson, the bivariate Poisson and Poisson difference or Skellam distribution) that can be used for football data can be also found in the same book; see Examples 8.3–8.5 and at the book's webpage[4] for the relevant WinBUGS code.

Currently there is a wide variety of R packages that are available for analyzing football data starting from the footbayes package[5] developed by Prof. Leonardo Egidi (the first author of this book) and described in detail in Chapters 4 and 5 of this book. footBayes consists of functions for fitting widely known soccer models (double Poisson, bivariate Poisson, Skellam, Student's t) through Hamiltonian Monte Carlo and Maximum Likelihood estimation approaches using Stan. The package also provides tools for visualizing team strengths and predicting match outcomes.

Other packages that are currently active in CRAN can be found in

[4]`http://www.stat-athens.aueb.gr/~jbn/winbugs_book/home.html`
[5]`https://github.com/LeoEgidi/footBayes`

`https://cran.r-project.org/web/views/SportsAnalytics.html`. A comprehensive list (updated on 22/1/2025) of football/soccer R packages is the following

- goalmodel R package: available at `https://github.com/opisthokonta/goalmodel`. The goal model package is designed to fit models including the negative binomial, Conway-Maxwell-Poisson model and the ones of Dixon and Coles (1997) and Rue and Salvesen (2000) that predict the number of goals scored in sports matches.
- socceR which provides functions for evaluating soccer predictions and simulating results from soccer matches and tournament.
- ggsoccer which provides functions for visualizing soccer event data in ggplot2.
- regista R package implements the Dixon and Coles (1997) approach and models for expected goals (xG) among others; available at `https://github.com/Torvaney/regista`.
- soccerAnimate: an R package to create 2D soccer animations; available at `https://www.datofutbol.cl/soccer-animate-r-package/`
- ffanalytics R package for Fantasy Football Data Analysis; available at `https://fantasyfootballanalytics.net/2016/06/ffanalytics-r-package-fantasy-football-data-analysis.html`.

Moreover, the following R packages provide a variety of **football datasets**:

- bundesligR contains all final standings of the Bundesliga in Germany from 1964 to 2016.
- engsoccerdata: English and European Soccer Results 1871-2020; available at `https://github.com/jalapic/engsoccerdata`
- EUfootball provides European football match results for top leagues in England, France, Germany, Italy, Spain, Netherlands, and Turkey from 2010-2011 to 2019-2020.
- footballpenaltiesBL contains data and plotting functions for analyzing penalty kicks in the German Men's Bundesliga from 1963-64 to 2016-17.
- footballR: R package to obtain football (soccer) data from APIs; available at `https://github.com/dashee87/footballR`.
- FPLdata contains functions for retrieving player attributes on Fantasy Premier League.
- worldfootballR: An R Package to Extract World Football (Soccer) data from `Fbref.com` and `transfermarkt.com`; available at `https://github.com/JaseZiv/worldfootballR`.

2.8 Basic model assumptions and model checking issues

In order to be able to use the double Poisson vanilla model, as described in Sections 2.1.1 and 2.1.2, we need to make the assumptions of

1. Independence of the goals scored by the two opponents.
2. The variance is equal to the expected goals (there is no under or over-dispersion).
3. There is no excess of draws in the data in comparison to the predicted ones.
4. The team abilities are constant across the season/tournament.

Nevertheless, during the last 25 years, the development of football analytics models and the related empirical studies have emerged a variety of objections concerning the above assumptions or characteristics of the standard double Poisson model. Hence, a modern football analytics modeler needs to check in his dataset whether the above assumptions hold and, in case that these assumptions are not effectively reflecting the characteristics of a football league or competition, to extend appropriately the model and embody the above characteristics in their predictive approach. Hence, the main questions (associated to the above assumptions) that a modern football analyst needs to asses are four:

1. Is there dependence or independence between the goals of two opponents?
2. Is there over-dispersion (or under-dispersion) in the number of goals scored?
3. Is there an excess of draws in the data in comparison to the model predictions?
4. Are the team abilities dynamic? If covariates are used instead, do we need to consider any dynamic effects?

2.8.1 Dependence in the number of goals

Naturally, the number of goals scored by the two opponents in a game should be dependent. In practice, small (but significant) dependence has been observed, as remarked in Karlis and Ntzoufras (2003) and McHale and Scarf (2011b). A number of models has been introduced in the literature in order to account for the plausible dependence between the goals scored by each team. The first model in the literature dealing with the dependence between the goals scored in a game is the bivariate Poisson model of Karlis and Ntzoufras (2003). This was followed by the Skellam model (Karlis and Ntzoufras, 2009) used for the difference which eliminates any linear correlation between the goals scored by the two opponent teams in game. These publications were followed by Koopman and Lit (2015, 2019a); Smit et al. (2020).

Another way to deal with positive dependence of the game goals is to consider the usual normal random effects (or Bayesian hierarchical model) in order to account for specific type of association, as in Baio and Blangiardo (2010). Nevertheless, such models do not seem to fit very well the football data of several leagues.

Finally, copula models seems to be the promising alternative for modelling complicated association structures in football. Nevertheless, estimation of copula models for discrete models have several estimation problems reported in the bibliography (Ötting and Karlis, 2023).

Empirically, slight correlation of the magnitude of 0.3 has been reported in the literature but this seems to diminish over the last years; estimated correlation using EUfootball package data is found to be slightly negative (-0.08) for data for a period of ten years (2010-2020) and seven popular leagues.

Finally, a theory for the observed small but significant correlation is that the distribution of the corresponding estimator might be bimodal reflecting two latent groups of games with (a) positive correlation (usually between teams of similar strength), and (b) negative correlation (usually in game with a clearly dominant team). Nevertheless, this theory has not been checked or confirmed in practice.

2.8.2 Over-dispersion

There is a variety of publications reporting slight by significant over-dispersion in the football analytics literature (Baxter and Stevenson, 1988; Karlis and Ntzoufras, 2000b).

To account of this we can use the negative binomial model or the generalized Poisson model (Ntzoufras, 2011, Section 8.3.2) and in Chapter 4 of this book.

Also, the usual normal random effects models (or the corresponding Bayesian hierarchical models) may account for over-dispersion (Baio and Blangiardo, 2010), (Ntzoufras, 2011, Section 9.3.1) although it seems that such models are much worse than the negative binomial or the generalized Poisson model.

On the other hand, under-dispersion is more rarely reported for football data. Moreover, discrete distribution for handling under-dispersion are not so common. The generalized Poisson, discussed above, can partially handle under-dispersed data but it does not seem to have been used for football data. Another distribution is the Poisson-COM (or Conway-Maxwell Poisson) distribution which was proven promising in the implementation on Premier League data (where some leagues with under-dispersed parameters were identified). Another implementation of a variety of distributions handling under-dispersion on several Premier league and Bundesliga seasons was presented in `http://opisthokonta.net/?p=1210`. In this analysis, five different distributions were used: the negative binomial, the Poisson-COM, the double-Poisson (it should not be confused with the double Poisson model used here), the Poisson-inverse Gaussian and the Delaporte distributions. In all 10 leagues, the fitted models report under-dispersion but in a vanilla type model without a constant parameter (which assumes an expectation of one goal per game for the guest team).

2.8.3 Excess of draws

Usually goal-based models do not fit or predict very well the draws. In practice we have an excess of draws in our data in comparison to the ones predicted from the goal/score-based models. This is more intense for the 0-0 and 1-1 draws, which are two of the most frequent scores in football. An easy remedy to overcome this problem is to include an additional draw inflation component in our modelling formulation. Nevertheless, some exceptions have been reported in the English Premier league where in some seasons the goal-based models over-estimate the number of draws.

From the early years of statistical analysis of football data, Dixon and Coles (1997) have foreseen the need to include extra parameters in order to estimate accurately the probability of a draw. In their seminal publication, they have considered an extra parameter in order to improve the fit of 0-0 and 1-1 draws which are two of the most frequent results in football. Their approach was followed by Karlis and Ntzoufras (2003) who added a diagonal inflated component in their bivariate Poisson model in order to capture the excess of draws. Similarly, Karlis and Ntzoufras (2009) have used a zero-inflated component in their Skellam based model for the score difference.

After these publications, several authors have underlined the need of extra model component in order to accurately estimated the probability of a draw.

The problem actually arises from the fact that when teams are of equal strength, then the expected number of goals should be identical while any probability of a draw is reasonable. So in practice the probability of a draw is not clearly identified. Nevertheless, when using specific distributions such as the Poisson distribution, the probability of a draw is automatically specified by the expected value (and other parameters); see for example Figure 2.5 which depicts how the probability of the draw changes with equal expected number of goals under the Poisson assumption.

As you can see from this figure, the probability of a draw is as high as 80% for teams with very low expected scoring ability and decreases around to 20% for teams with expected number of goals equal to three (which is quite high and unrealistic for most games). Although the principle that the highest are the expected scoring abilities of the equal teams, the lower is the probability of observing a draw seems to be reasonable, the rate of increase does not seem to reflect the reality and some extra parameters are needed in order to capture this important characteristic of football.

2.8.4 Dynamic abilities

Another important aspect of the vanilla models is the assumption of "fixed" (or static) attacking and defensive team abilities. By fixed we mean that the team abilities in model (2.3) of Section 2.1.2 are assumed to be constant across time. Nevertheless, football fans intuitively believe that each team has ups

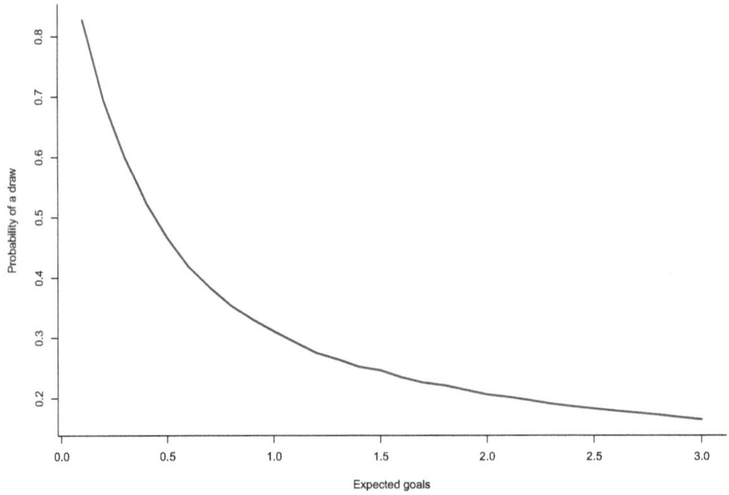

FIGURE 2.5
Probability of a draw for teams of equal scoring rate under the Poisson distribution.

and downs in their performance which should be reflected in their attacking and defensive abilities. Moreover, in most sports an underlying psychological factor is assumed which creates the expectation that when a team wins or over-performs (i.e. achieves a result which is better than what was expected), then this causes a positive team mood which eventually might lead to good performances and sequential successful results. This time dependence of the abilities is usually modelled by introducing simple random effects in the model formulation or by adapting a hierarchical structure with the Bayesian framework. These random ability parameters are usually assumed to have a random walk behaviour. Such time dependent abilities are called in the bibliography *dynamic team abilities* (Owen, 2011; Koopman and Lit, 2015) and they reflect the performance of the teams in a more enhanced way than simple fixed ability models (Egidi et al., 2018b).

Concerning the formulation, the attacking and defensive abilities att_k and def_k in (2.3) are substituted by $att_{k,t}$ and $def_{k,t}$, respectively. The sub-script t is introduced to the team parameters in order to denote the team ability of the k-th team at game/week t; where $t \in 2, \dots, T$ and $T = 2K - 2$. Moreover, the time dependent abilities are incorporated in the model formulation through the equations:

$$att_{k,t} \sim N(att_{k,t-1}, \sigma_a^2) \text{ and } def_{k,t} \sim N(def_{k,t-1}, \sigma_d^2).$$

Note that this is the simplest time dependence structure we may consider. Although other dynamic structures can be considered, this is by far the most popular approach to account for time-varying team performance. Overall, the dynamic abilities model is by far more realistic than the fixed effects model. Of course, this comes to a price. The dynamic model is more computationally demanding in terms of fit. For this reason, many researchers may consider to use the simpler fixed effects formulation in order to obtain faster results, especially if evidence from model comparison measures does not suggest important improvement of the fit (or the prediction). In any occasion, a dynamic model might be valuable in terms of interpretation of the team performance across the season(s).

2.9 How to compare and select models: Criteria, assumptions

In this section, we frequently summarise the possible alternatives for comparing football analytics models. This is standard theory within statistical methods and it is summarized here in order to have a complete picture of standard and more modern ways of model comparison and evaluation. In this section we focus on the main response (goals or win/draw/loss), while in subsequent sections we also extend the approach to other measures of interest like the number of points earned in the league or the final predicted league ranking.

2.9.1 Goodness of fit and significance tests

The double Poisson models described in Sections 2.1.2 and 2.2.3 and the corresponding multinomial models that can be used for the match outcome fall in the class of generalized linear models (GLM) while the random effects models described in Section 2.8.4 fall in the context of generalized linear mixed models (GLMM). Hence, we may use the standard approaches within this theory to check for certain hypotheses or assumptions. We refer the reader to use the excellent books of Alan Agresti for GLMs and categorical data (Agresti, 2013, 2015) as technical companions to this book; especially for readers interested to go deeper in the relevant statistical theory.

Within GLM framework we can use standard Wald t-tests (Wald, 1943) and likelihood ratio tests χ^2 significance tests using the model deviance measures (Wilks, 1938) for assessing the significance of each added factor and ability parameter.

Wald tests are readily available from the output of any statistical software including the glm function in R. They are used to test the hypothesis that a specific parameter β_j of a GLM is equal to zero or not (see model in Section 2.2.3 for our context) and hence if the corresponding covariate should be included or not in the model. Hence it tests for the *null* hypothesis $H_0 = \beta_j = 0$ vs. the *alternative* $H_1 = \beta_j \neq 0$ using the simple test statistic

$$Z_{\beta_j} = \frac{\widehat{\beta}_j}{se(\widehat{\beta}_j)},$$

which is assumed to follow asymptotically the normal distribution with mean zero and variance equal to one, denoted as $N(0,1)$, and referred as the standardized normal distribution; where $\widehat{\beta}_j$ is the maximum likelihood estimator of a coefficient and $se(\widehat{\beta}_j)$ its corresponding standard error; all these measures and the corresponding p-value are readily available in R in the summary of the output of glm function.

Note that testing for differences between team abilities is not of major interest in football modelling prediction. The main reason to consider such an analysis is for interpretation purposes where a clustering of the teams in terms of performance might be of interest. In the case we implement multiple testing on defensive and attacking parameters, then it will be reasonable to also apply multiple testing corrections.

Likelihood ratio tests (LRTs) comprise a more general framework for comparing two nested models $M_0 \subset M_1$. A model M_0 is nested to a model M_1 when we can obtain it by restricting a subset of parameters of M_1 to specific values. So if we assume that model M_0 has parameters $\boldsymbol{\theta}_0 = \boldsymbol{\vartheta}_0$ and model M_1 has parameters $\boldsymbol{\theta}_1 = (\boldsymbol{\vartheta}_0, \boldsymbol{\vartheta}_1)$, then the LRT will test for $H_0 : \boldsymbol{\vartheta}_1 = 0$ vs. $H_1 : \boldsymbol{\vartheta}_1 \neq 0$ which can be interpreted that the two models are identical in terms of fit (H_0) vs. model M_1 is better in terms of fit.

The LRT is based on the notion of the deviance of model M with parameters θ_M which is given by the model maximized log-likelihood $f(\boldsymbol{y}|\boldsymbol{\theta}_M, M)$ multiplied by the factor of minus two (-2). The model deviance is given by

$$D_M = -2 \log f(\boldsymbol{y}|\widehat{\boldsymbol{\theta}}_M, M) \quad \text{with} \quad \log f(\boldsymbol{y}|\widehat{\boldsymbol{\theta}}_M, M) = \sum_{i=1}^{n} \log f(y_i|\widehat{\boldsymbol{\theta}}_M, M),$$

where $f(y_i|\widehat{\boldsymbol{\theta}}_M, M)$ is the probability density function of model distribution adopted in our model formulation. For the Poisson-based model, the sampling density is given by

$$\log f(y_i|\lambda_i, M) = -\lambda_i + y_i \log(\lambda_i) - \log(y_i!)$$

while for the multinomial models is given by

$$\log f(y_i|p_{i,1}, p_{i,2}, p_{i,2}, M) = \sum_{k=1}^{3} \mathcal{I}(y_i = k) \log p_{i,k},$$

where $\mathcal{I}(A)$ is the indicator function taking the value of one when A is true and zero otherwise, y_i is the final game outcome taking values 1, 2 or 3 for win, draw and loss of the home team and $p_{i,k}$ are the corresponding outcome probabilities. As you may understand, for a sequence of nested models $M_0 \subset M_1, \ldots, M_K$ the deviance measure is a decreasing function of the model dimension. The more complicated the model is, the lower is the deviance.

The LRT is based on the finding of Wilks (1938) that the deviance difference $D_0 - D_1$ of two nested models $M_0 \subset M_1$, (with p_0 and p_1 parameters, respectively) asymptotically follows the χ^2 with degrees of freedoms equal to the difference of dimensions of the models under comparison, $p_1 - p_0$.

Under this perspective, in the general GLM framework a standard model to start with is the constant (or null model) where the same parameter is used for all observations. Under this perspective, in the Poisson framework for football data, it would correspond to consider a Poisson model with parameter $\lambda = \overline{Y} = (\overline{Y}_1 + \overline{Y}_2)/2$, i.e. the overall mean of goals scored by both home and away teams. Since the home effect is well established and acceptable in the football literature, we would suggest to consider different parameters for home and away games given by the means $\overline{Y}_1$ and $\overline{Y}_2$ of the goals of home and away teams. Moreover, in a second level comparison, each new model should be compared with the standard vanilla model (see Equation 2.3) since this model is well established in the literature as a well behaved model which provides acceptable fit and prediction.

In terms of **overall goodness of fit**, a simple χ^2 test can be used comparing the fitted model with the "saturated" model. As "saturated" model we define the model which has expected values equal to the observed values, then, with as many parameters as observations. This definition is problematic in our framework when using either the goals or the final outcome as response. For the first case, some goals will be equal to zero so a Poisson model with such a correlation will not be able to be fitted. A simplified approach is for the cases where zero goals are observed to consider as the mean of the Poisson an arbitrary small value λ_0, say equal to 0.1. Nevertheless, all results will strongly depend on this value. A slightly better alternative is to estimate this λ_0 by using the mean of the fitted values of the vanilla model for all cases with zero observations. All these modifications do not ensure that the distribution of the deviance difference will still follow a χ^2 distribution, so using bootstrap is suggested to estimate the induced significance.

The problem is more severe in the case of binomial or multinomial modelling when we consider the game outcome (home win, draw or away win) as response. In such cases, the test proposed by Hosmer and Lemeshow (1980) for binary logistic regression can be used instead. The logic of this test is simple and based on the formalization of $g = 10$ groups based on the quantilies of the predicted probability of π. Then the sum of the success probabilities for each group is an estimate of the expected frequencies e_k for each group k and this is compared with the corresponding observed successes o_k using a simple Pearson's χ^2 test. Under this approach the two competing hypotheses can be written as

H_0: there is no difference between observed and expected frequencies
vs.
H_1: there are differences between observed and expected frequencies for each group $k = 1, \ldots, g$.

Under this approach, the test statistic is given by

$$X_{HL} = \sum_{k=1}^{g} \frac{(o_k - e_k)^2}{e_k(1 - e_k/N_k)}$$

with

$$
\begin{aligned}
o_k &= \sum_{i=1}^{n} \mathcal{I}(\widehat{\pi}_i \in \mathcal{Q}_k) y_i \\
e_k &= \sum_{i=1}^{n} \mathcal{I}(\widehat{\pi}_i \in \mathcal{Q}_k) \widehat{\pi}_i \\
N_k &= \sum_{i=1}^{n} \mathcal{I}(\widehat{\pi}_i \in \mathcal{Q}_k).
\end{aligned}
$$

In the above equations, $\widehat{\pi}_i$ are the fitted success probabilities under the assumed model; $\mathcal{Q}_k$ is the interval $(q_{k-1}, q_k]$ with q_k denoting the k-th quantile of $\widehat{\pi}_i$ and with $q_0 = 0$ and $q_g = 1$; f_k and e_k are the observed and expected successes for observations with $\widehat{\pi}_i$ in the interval $\mathcal{Q}_k = (q_{k-1}, q_k]$; N_k are the corresponding number of observations in $\mathcal{Q}_k$. The above test statistic X_{HL} follows a χ^2_{g-2} distribution, with $g-2$ degrees of freedom. The test can be implemented using the R function hoslem available in the package ResourceSelection. Alternatively, the package generalhoslem offers a generalized version of the test; see function logitgof.

For our case, the generalization of Hosmer and Lemeshow test for multinomial data (Fagerland et al., 2008) can be used instead. For the football data, this will be modified to

$$
X_{GHL} = \sum_{k=1}^{g} \sum_{c=1}^{2} \frac{(o_{k,c} - e_{k,c})^2}{e_{k,c}}
$$

with

$$
\begin{aligned}
o_{k,c} &= \sum_{i=1}^{n} \mathcal{I}(\widehat{\pi}_i \in \mathcal{Q}_k) y_{i,c} \\
e_{k,c} &= \sum_{i=1}^{n} \mathcal{I}(\widehat{\pi}_i \in \mathcal{Q}_k) \widehat{\pi}_{i,c},
\end{aligned}
$$

where $o_{k,c}$ and $e_{k,c}$ are the observed and expected frequencies of the c outcome. For the formation of the group intervals, the probability $\widehat{\pi}_i = 1 - \widehat{\pi}_{i1}$ is used, following the suggestion of Fagerland et al. (2008). The generalized Hosmer and Lemeshow test statistic under the null hypothesis follows the χ^2 distribution with $(g-2) \times (c-1)$ degrees of freedom – in our case $df = 2g - 4$. The generalized Hosmer and Lemeshow test for multinomial data can be implemented via the R function logitgof available in the package generalhoslem.

Before we close this section, we should state that this test is sensitive on the choice of the number of groups g. Moreover, it fails to identify certain types of lack of fit since it refers to grouped data and marginal comparisons in terms of fit (Hosmer et al., 1997).

2.9.2 Model comparison using information criteria

The use of information criteria in model choice was introduced in the early seventies in order to find a consistent method for model selection. The most popular criteria are Akaike's Information Criterion (AIC; Akaike, 1973), and the Bayes Information Criterion (BIC; Schwarz, 1978). These criteria have been widely used by all statisticians and modern machine learners although both BIC and AIC have derived from Bayesian arguments. Specifically, BIC is an approximation of the Bayes factor (see Section 2.9.4) used for Bayesian model comparison under specific conditions; see for details in Schwarz (1978) and Kass and Wasserman (1995). On the other hand AIC was derived as an approximately unbiased estimate of the Kullback-Leibler discrepancy between the model under consideration and the true model formulation (Akaike, 1973).

Generally, most information criteria suggest the selection of the model which minimize a penalized deviance measure (or, roughly speaking, it is based a penalized maximum log-likelihood measure) given by

$$IC_m = -2\log\Big(f(\boldsymbol{y}|\hat{\boldsymbol{\theta}}_m, m)\Big) + d(m)F, \tag{2.8}$$

where $\boldsymbol{\theta}_m$ is the whole parameter vector, $\hat{\boldsymbol{\theta}}_m$ are the corresponding maximum likelihood estimates, and $d(m)$ is the dimension of the parameter vector $\boldsymbol{\theta}_{(m)}$.

In linear regression models $\boldsymbol{\theta}_m^T = [\boldsymbol{\beta}_{(m)}^T, \sigma^2]$ and minimizing $-2\log\Big(f(\boldsymbol{y}|\hat{\theta}_m, m)\Big)$ is equivalent to minimizing $n\log(RSS_m)$. Also note that F is the penalty imposed to the -2log-likelihood for each additional parameter used in the model. Different penalty functions result in different criteria; for example for $F = 2$ we have AIC and for $F = \log(n)$ we have BIC.

If we want to compare two models M_0 and M_1 then we select the one that has lower value of IC and therefore we define as IC_{01} the difference of the two information criteria. Hence

$$IC_{01} = -2\log\left(\frac{f(\boldsymbol{y}|\hat{\theta}_{M_0}, M_0)}{f(\boldsymbol{y}|\hat{\theta}_{M_1}, M_1)}\right) - [d(M_1) - d(M_0)]F. \tag{2.9}$$

Without loss of generality, we assume that $d(M_0) < d(M_1)$. Note that if $IC_{01} < 0$ we select model M_0 and if $IC_{01} > 0$ we select model M_1. We can generalize the above criterion difference by substituting the expression $[d(M_1) - d(M_0)]F$ by a more complicated penalty function ψ. In such case we may write the information criteria in more general setup given by

$$IC_{01} = -2\log\left(\frac{f(\boldsymbol{y}|\hat{\boldsymbol{\theta}}_{M_0}, M_0)}{f(\boldsymbol{y}|\hat{\boldsymbol{\theta}}_{M_1}, M_1)}\right) - \psi,$$

where ψ is a penalty function depending on difference of the model dimensions, $d(M_1) - d(M_0)$, sample size n, and design matrices, $\boldsymbol{X}_{(M_0)}$ and $\boldsymbol{X}_{(M_1)}$.

Shao (1997) divides model choice criteria in three major divisions:

1. Asymptotically valid criteria under the assumption that a true model exists.
2. Asymptotically valid criteria under the assumption that not a fixed dimension true model exists.
3. A compromise between these two categories.

The main conclusion of Shao (1997) was that IC_m with $F = 2$ and $F \to \infty$ when $n \to \infty$ are two differently behaved categories of criteria referred as AIC-like and BIC-like criteria. The BIC-like criteria perform better if the true model has simple structure ("finite dimension") while the AIC-like criteria are better if the true model is a more complex one ("infinite dimension"). The main argument of Zhang (1997) in favour of BIC-like criteria is that the existence of a true model is doubtful and even if exists we may prefer to select a simpler model that approximates sufficiently the true one. In his words, "the practical advantage of a more parsimonious model often overshadows concerns over the correctness of the model. After all the goal of statistical analysis is to extract information rather to identify the correct model." In this direction, Rissanen (1986) states that it is obvious that all selection criteria give rise to quantification of the parsimony principle. They differ in the weight (or significance) that they give to goodness of fit and model complexity. The goodness of fit is measured by the log-likelihood ratio while model complexity by the number of model parameters.

Another interesting perspective of AIC is that "leave-one-out" cross-validation method is asymptotically equivalent to AIC and C_p, as noted by Stone (1977) and Shao (1993).

2.9.3 Bayesian predictive measures

The criteria introduced in Section 2.9.2 serve as useful tools for model comparisons in a classical statistical framework, however they are not well suited for capturing models' uncertainty when framed in a Bayesian perspective. In fact, none of the above method uses any form of parameters' or model uncertainty to derive a final measure of predictive fit: the predictive accuracy is somehow measured through the pointwise log-likelihood evaluated in the maximum likelihood plugin estimate, by not capturing any posterior uncertainty and using just a rough measure of goodness of fit. Not only: in the criteria above, the bias correction uses the "nomina" number of parameters, with the consequence of a possible over-penalization in sparse or hierarchical models exhibiting a large amount of shrinkage—think, for instance at the case where only a bunch of covariates are statistical significant, with the majority of the parameters set at zero; or when in a hierarchical models the group-parameters come from an exchangeable prior distribution, which makes many of them *a posteriori* overlapped and not distinguishable from the others. For such a reason, the number of nominal parameters in the model could be far from being the "true" number of parameters.

The criteria introduced in this section try to overcome these issues by:

- Replacing the log-likelihood conditioned on the MLE with more relevant Bayesian tools (WAIC and LOOIC);
- Treating the number of parameters as a random variable and trying to estimate the *effective* number of parameters through the uncertainty arising from the posterior distribution (DIC and WAIC);
- Actually considering the predictive distribution of future observable outcomes (as in WAIC and LOOIC).

2.9.3.1 DIC

The Deviance Information Criterion (DIC) proposed by Spiegelhalter et al. (2002) is a Bayesian version of AIC in Equation (2.8) and makes two changes, by replacing the maximum likelihood estimate $\hat{\boldsymbol{\theta}}_m$ with the posterior mean $\hat{\boldsymbol{\theta}}_{m,\text{Bayes}}$ and the number of parameters $d(m)$ with a data-based estimation of the effective number of parameters, $\tilde{d}(m)$, in such a way that the criterion for model m is now:

$$DIC_m = -2\log\left(f(\boldsymbol{y}|\hat{\boldsymbol{\theta}}_{m,\text{Bayes}}, m)\right) + 2\tilde{d}(m). \tag{2.10}$$

There are two possible approaches to estimate the effective number of parameters, where according to the first one this number is defined as:

$$\tilde{d}(m) = \log(f(\boldsymbol{y}|\hat{\boldsymbol{\theta}}_{m,\text{Bayes}}, m)) - \text{E}_{\text{post}}(\log(f(\boldsymbol{y}|\boldsymbol{\theta}_m, m)), \tag{2.11}$$

where the expectation in the second term is the average of $\boldsymbol{\theta}$ over its posterior distribution. Technically, Equation (2.11) is computed using the MCMC simulations from the posterior distribution as introduced in Section 2.5.1. Alternatively, the effective number of parameters could be defined as:

$$\tilde{d}(m) = \text{var}_{\text{post}}(\log(f(\boldsymbol{y}|\boldsymbol{\theta}_m, m)), \tag{2.12}$$

where the variance is computed over the posterior distribution of the parameter vector $\boldsymbol{\theta}_m$. Both the expression (2.11) and (2.12) provide the correct answer in the limit of fixed model and large n; for linear models with uniform prior distributions, both these quantities converge to the number of parameters $d(m)$. Note that, in practical contexts, $1 \leq \tilde{d}_m \leq d(m)$.

2.9.3.2 WAIC

Another recent and popular criterion is the Watanabe-Akaike Information Criterion (WAIC) introduced by Watanabe and Opper (2010), a fully Bayesian approach specifically designed to measure the out-of-sample predictive accuracy. In order to fulfil this task, the criterion replaces the log-likelihood in the previous criteria with the log-pointwise predictive density for model m is

$$\text{lppd}_m = \sum_{i=1}^{n} \log \left(\int f(y_i|\boldsymbol{\theta}_m, m) f(\boldsymbol{\theta}_m|y_i) d\boldsymbol{\theta} \right). \tag{2.13}$$

Similarly as in the DIC formulation, there is a correction consisting on the effective number of parameters, estimated either as:

$$\tilde{d}(m) = \sum_{i=1}^{n} (\log(\text{E}_{\text{post}}(f(y_i|\boldsymbol{\theta}_m, m)) - \text{E}_{\text{post}}(\log(f(y_i|\boldsymbol{\theta}_m)))), \tag{2.14}$$

or by using the variance of individual terms in the log predictive density and taking the sum over the n data points:

$$\tilde{d}(m) = \sum_{i=1}^{n} \text{var}_{\text{post}}(\log(f(y_i|\boldsymbol{\theta})), \tag{2.15}$$

where this expression is similar to Equation (2.12), with the difference that Equation (2.15) is more stable because it computes the he variance separately for each data point and then sums. WAIC is then defined as

$$WAIC_m = -2\text{lppd}_m + 2\tilde{d}(m), \tag{2.16}$$

by using either Equation (2.14) or (2.15) as a bias correction. Similarly as for Equations (2.11) and (2.12), the $\tilde{d}(m)$ in either Equation (2.14) or (2.15) could be computed from simulations by replacing the expectations by averages over the MCMC posterior draws.

As motivated by Gelman et al. (2014), compared to other information criteria such as AIC and DIC, WAIC averages over the posterior distribution rather than conditioning on a point estimate, such as the maximum likelihood estimate or the posterior mean. This feature makes WAIC appealing from a predictive perspective, since it evaluates the posterior predictive predictions: AIC, BIC, and DIC estimate the predictive performance by using a plugin density, but the Bayesian community would still use the posterior predictive density for future observable outcomes, in place of some plugin surrogates. Moreover, WAIC is particularly relevant for hierarchical and mixture models in which the number of parameters increases with sample size and for which point estimates often are not useful.

2.9.3.3 LOOIC

Another way to assess and compute predictive accuracy is based on leave-one-out cross-validation (LOO) (Vehtari et al., 2017), a method for estimating pointwise out-of-sample prediction accuracy from a fitted Bayesian model using the log-likelihood evaluated at the posterior simulations of the parameter values. When we perform Bayesian cross-validation, we repeatedly partition the data into a training set and a holdout set, and then the model is fit and the posterior distribution is obtained from training data; this fit is then evaluated from a predictive perspective using an estimate of the log predictive density of the holdout data,

$$\text{lppd}_{\text{loo-cv},m} = \sum_{i=1}^{n} \log(f(y_i|y_{-i})) = \sum_{i=1}^{n} \log(\int f(y_i|\boldsymbol{\theta}_m, m) f(\boldsymbol{\theta}_m|y_{-i}) d\boldsymbol{\theta}), \tag{2.17}$$

where $f(y_i|y_{-i})$ is the leave-one-out predictive density given the data without the i-th data point. The main difference with the lppd defined in (2.13) for the WAIC in the previous section is that cross-validation uses every time $n-1$ data to fit the model and one residual holdout point to test the predictive accuracy. To effectively compute $\text{lppd}_{\text{loo-cv}}$ one could use the MCMC simulations from the posterior distribution, as explained by Gelman et al. (2014) and Vehtari et al. (2017). Moreover, to adequately evaluate the posterior predictive distribution $f(y_i|y_{-i})$ when the n points are conditionally independent in the data model, Vehtari et al. (2017) suggest to use importance sampling ratios, especially the Pareto smoothed importance sampling (PSIS) technique (Vehtari et al., 2015).

Analogously as for the other criteria, the LOOIC is then defined by scaling a measure of predictive accuracy for the factor -2:

$$LOOIC_m = -2\text{lppd}_{\text{loo-cv},m}. \tag{2.18}$$

Note that in Equation (2.18) none bias correction applies: however, an effective number of parameter estimation can be obtained under LOOIC as well.

Analogously as with the other predictive information criteria such as AIC, BIC, DIC and WAIC, the lower is the LOOIC, the better is the model predictive accuracy. The rationale behind the use LOO or WAIC is to estimate the accuracy of the predictive distribution by requiring data to be divided into disjoint, ideally conditionally independent, pieces. This feature makes LOOIC and WAIC much more appealing from a Bayesian predictive perspective, however this data distinction could represent a limitation when applied to structured, for instance hierarchical, models. In addition, cross-validation methods can be computationally expensive unless we are framed in settings where some shortcuts are available to approximate the distributions $f(y_i|y_{-i})$ without re-fitting the model model each time.

2.9.4 Bayesian model comparison and variable selection

The Bayesian approach is quite different than the classical approach in the way that it treats both hypothesis tests and model comparison. Essentially, the main approach assumes that every hypothesis corresponds to a model and therefore hypothesis tests are simply referring to the comparison of pairs of models. There is no requirement for the two models under comparison to be nested although hypothesis tests are more naturally framed in the nested model comparison setup.

Another characteristic here is that the model comparison on the classical approach usually relies in the maximum likelihood ratio (i.e. checks how much the fit changes via a comparison of the maximum likelihoods of the two competing models) while in the Bayesian approach we base our decision in the comparison of "averaged" marginal likelihoods of each model.

2.9.4.1 Bayes factor and posterior model odds

In order to proceed with Bayesian hypothesis tests and/or model comparison, we need to expand our model formulation by considering m to be an model indicator which will be treated as a (discrete) parameter under estimation. Under this approach, we need to attach a prior model probability $f(m)$ and then calculate the posterior model probability in a similar manner as in the single model approach described in Section 2.5.

For the hypothesis test setup, we need to consider only two models, that is $m \in \{M_0, M_1\}$. So let us consider the case that we wish to compare two models M_1 and M_0 with parameters $\boldsymbol{\theta}_{M_1}$ and $\boldsymbol{\theta}_{M_0}$, respectively. Under this setup we need to estimate $(m, \boldsymbol{\theta}_{M_0}, \boldsymbol{\theta}_{M_1})$ from the following posterior distributions of interest

1. The posterior distribution of the parameters of model M_k: $f(\boldsymbol{\theta}_{M_k}|\boldsymbol{y})$ for $k = 0, 1$.
2. The posterior probabilities of the two models given by $f(M_k|\boldsymbol{y})$ for $k = 0, 1$.

For the model comparison between M_0 and M_1 we are interested for the comparison of $f(M_1|\boldsymbol{y})$ and $f(M_0|\boldsymbol{y})$ via the calculation of their ratio which is called posterior model odds and is given by

$$PO_{10} = \frac{f(M_1|\boldsymbol{y})}{f(M_0|\boldsymbol{y})} = \frac{f(\boldsymbol{y}|M_1)}{f(\boldsymbol{y}|M_0)} \times \frac{f(M_1)}{f(M_0)} = B_{10} \times \frac{f(M_1)}{f(M_0)}, \qquad (2.19)$$

which is said to be the *posterior model odds* of model M_1 versus model M_0.

The ratio of prior model probabilities $\frac{f(M_1)}{f(M_0)}$ defines prior preference towards one or another model. A naive approach is to assume that the prior model probabilities are equal when no other information is available. Under this approach, the posterior model odds is equal to the *Bayes factor* B_{10}.

The Bayes factor of model M_1 versus model M_0, B_{10}, is defined as the ratio of the "marginal" likelihoods $f(\boldsymbol{y}|M_1)$ and $f(\boldsymbol{y}|M_0)$. The Bayes factor can be expressed as the posterior odds divided by the prior odds of two compared models. Therefore, it quantifies the prior to posterior change of relative evidence for the two compared models.

The marginal likelihood of any model M is the model likelihood "averaged" over the prior distribution. It is also called the Bayesian likelihood or the Bayesian "evidence" and it is given by

$$f(\boldsymbol{y}|M) = \int f(\boldsymbol{y}|\boldsymbol{\theta}_M, M) f(\boldsymbol{\theta}_M|M) d\boldsymbol{\theta}_M. \qquad (2.20)$$

Integral (2.20) is analytically tractable when the conjugate prior approach is adopted. For other priors, numerical or MCMC methods must be used instead.

Posterior model odds PO_{10} (and Bayes factors B_{10}) allow for a straightforward Bayesian model comparison and testing according to Jeffreys' interpretation. Under his suggestion, for $PO_{10} > 1$ (or $B_{10} > 1$) we have evidence in favour of model M_1 which is characterized as:

- *negligible* or "not worth than a bare mention" for $PO_{10} \in (1, 3]$,
- *positive* for $PO_{10} \in (3, 20]$,
- *strong* for $PO_{10} \in (20, 150]$,
- *very strong* for $PO_{10} > 150$.

For $PO_{10} < 1$ we have evidence in favour of the null model using similar interpretation as for $PO_{01} = 1/PO_{10}$; for more details, see (Kass and Raftery, 1995).

The posterior probabilities can be directly obtained from posterior model odds using the expression

$$f(m_k|\boldsymbol{y}) = \frac{PO_{k0}}{\sum\limits_{m_\ell \in \mathcal{M}} PO_{\ell\,0}} = \frac{1}{\sum\limits_{m_\ell \in \mathcal{M}} PO_{\ell k}} \,.$$

The above pairwise comparison can be easily extended to a model comparison setup where $m \in \mathcal{M}$; where $\mathcal{M}$ is the set of models under comparison, commonly called model space. So now the focus is given on *posterior model probabilities* or *weights* $f(m|\boldsymbol{y})$ given by

$$f(m|\boldsymbol{y}) = \frac{f(\boldsymbol{y}|m)f(m)}{\sum_{m' \in \mathcal{M}} f(\boldsymbol{y}|m')f(m')} \,. \tag{2.21}$$

2.9.4.2 Posterior probability of variable inclusion

In variable selection literature, the model indicator m is usually substituted by a vector of binary indicators $\boldsymbol{\gamma}$ of size p. Each γ_j corresponds to β_j with $\gamma_j = 1$ if X_j is included in the model (i.e. $\beta_j \neq 0$) and $\gamma_j = 0$ otherwise. We usually include the constant term in all models, hence $\gamma_0 = 1$ with prior probability equal to one.

As the size of the model space is given by $|\mathcal{M}| = 2^p$, even a moderate number of potential covariates results in a large number of models from which the best one has to be selected. For example for $p = 20$ covariates more than one million models have to be considered. In such cases, all posterior model weights will be low even if a small group of models is much better than the remaining ones. Alternatively, we may select a model based on the posterior inclusion probabilities

$$f(\gamma_j = 1|\boldsymbol{y}) = \sum_{\boldsymbol{\gamma}_{\setminus j} \in \{0,\,1\}^{p-1}} f(\gamma_j = 1, \boldsymbol{\gamma}_{\setminus j}|\boldsymbol{y}). \tag{2.22}$$

In practice, this probability is the sum of posterior probabilities for all models which include covariate X_j in their linear predictor.

2.9.4.3 Selection of models and covariates

Selection of a single model

If we wish to select a single "best" model then we choose the one with the maximum posterior probability $f(m|\boldsymbol{y})$ or identically $f(\boldsymbol{\gamma}|\boldsymbol{y})$. This model is called the *maximum a posteriori (MAP)*. Alternatively, we may use the posterior variable inclusion probabilities, to trace the *median probability (MP) model*, which is defined as the model with all covariates having $f(\gamma_j = 1|\boldsymbol{y}) > 0.5$. The MP model has better predictive performance than the MAP model under certain conditions; for details, see (Barbieri and Berger, 2004).

Reporting of a set of best models

An advantage of Bayesian model comparison is that we can evaluate posterior probabilities and hence also quantify the uncertainty concerning the best fitted models. If we wish to report a group of best models, following the suggestions of (Kass and Raftery, 1995), we may report models m_k that are similar in terms of posterior evidence to the MAP model. Hence we may restrict attention and report models with posterior model odds $PO_{MAP,k} < 3$, i.e. models which have a posterior probability that is a least 1/3 of the posterior probability of the MAP model.

Bayesian model Averaging

In certain cases we do not wish to select a specific model but rather want to make inference or predictions by taking into account model uncertainty in our analysis. Hence we may obtain the model averaged posterior density of any quantity of interest ξ by considering the posterior distributions $f(\xi|m, \boldsymbol{y})$ weighted by the corresponding model weights $f(m|\boldsymbol{y})$. The set of models we include in the model averaging procedure may be remarkably reduced by considering only the ones with $PO_{MAP,k} < 3$ or by including models with covariates having posterior inclusion probabilities higher than 0.5.

2.9.4.4 The "Paradox" in the room

Before closing this short section, we should mention an important problem of Bayesian model comparison: marginal likelihoods and the resulting Bayes factors are sensitive on the dispersion parameters of the priors $f(\boldsymbol{\beta}_M|M)$. This problem is widely known as the *Lindley–Bartlett* or *Jeffreys paradox* Lindley (1957); Bartlett (1957). Hence the specification of the prior parameters in variable selection problems becomes a very important issue which is partially diluted when using Zellner's g-prior setup (Zellner, 1986).

2.9.5 Training and testing our model

In modern statistics and data science, we need to train our model (i.e. estimate its parameters) on a "train" dataset. In order to test our model, in an out-of-sample fashion, we need to have an additional dataset called "test" or "validation" where we measure its accuracy using several measures. Most of the times, we have a single dataset which we split it in train and test sub-datasets in order to evaluate its prediction accuracy. Nevertheless, the biggest problem is how to split the dataset: what should be the size of each dataset and which observations should go in each segment. Usually random sampling is applied when observations are not depended of some kind. When the data are time dependent, then we should be careful with the selection of the train and the dataset since future data cannot be used to train a model for prediction in the past. Football data can be viewed as time series data since the data come in batches of weeks/game days. Hence, we will avoid random splitting even if the implemented model is not using temporal components.

Generally, when we wish to evaluate the predictive performance of the model, we split our dataset $\mathcal{D} = \{1, 2, \ldots, n\}$ in the train sub-sample $\mathcal{T}$ of size n^* and the test/validation dataset $\mathcal{V} = \mathcal{D} \backslash \mathcal{T}$ sub-sample of size $n_v = n - n^*$. In general contents, the split of the data is implemented randomly. Nevertheless, our data in football (and more generally in sports) are appearing sequentially every game day or week. Hence, the split of the season data arises naturally. Hence we usually have the data available up to week w^* (and $n^* = w^* K/2$) and we try to predict the final results for the rest of the matches. Therefore, n stands for the number of games of the full season, n^* the number of games available to the time point of interest and n_v the remaining, to be predicted, number of games while $\mathcal{T} = \{1, 2, \ldots, n^*\}$ and $\mathcal{V} = \{n^* + 1, 2, \ldots, n\}$.

Nevertheless, in the full regeneration of the league we will focus on the evaluation of goodness of fit of the model. In this occasion we do not split the dataset $\mathcal{D}$ but we consider the full dataset in order to both train and evaluate the (in-sample) performance of the model. Hence, we set $\mathcal{T} = \mathcal{V} = \mathcal{D}$.

2.9.6 Out-of-sample prediction

With the term *out-of-sample* (or *held-out*) prediction we mean the ability of a model to predict the final outcome (of a game) in future data which have not used in the learning or the estimation procedure. Hence we train the data using observations of data from a train dataset $\mathcal{T}$ and we test the predictive ability in the validation dataset $\mathcal{V}$. In order to be able to implement this out-of-sample prediction, we need to know the response or outcome values also in $\mathcal{V}$. This is not happening naturally when collecting data from experiments since $\mathcal{V}$ should refer to future observations where the outcome variable is unknown.

Hence the out-of-sample evaluation is usually implemented in the dataset at hand where the whole dataset $\mathcal{D}$ is split in $\mathcal{T}$ and $\mathcal{V}$. This procedure is called *cross-validation*. Naturally, many questions about the implementation of cross-validation arise. For example, how large shall we select the sample size n^* of train set and n_v of the validation set given the sample size n? How many times shall we repeat the procedure? For a random selection of splits or for all splits should we report the mean, standard deviation or the whole distribution of the metric we are considering?

A popular cross-validation approach is the k-fold cross-validation. In this approach we select the number of sub-datasets $\mathcal{D}_k$ that we are going to split the dataset via k, and indirectly the size of the validation set at each time which will be about $n_v = \lfloor n/k \rfloor$. Then we select the validation dataset as one of the sub-datasets i.e. $\mathcal{V} = \mathcal{D}_d$ and the train set to contain all observations by the rest of the sub-data, i.e. $\mathcal{T} = \mathcal{D} \setminus \mathcal{D}_d$. We repeat this procedure for $d = 1, \ldots, k$ and we calculate and report the mean of the evaluation measure. Usually, k is selected to be equal to ten. Another standard method is to select $k = n$ which results to the leave-one-out cross-validation approach.

2.9.7 Prediction evaluation metrics

For numeric variables, the most common metrics are:

- RMSE: the root mean square error given by $\sqrt{\frac{1}{n_v} \sum_{i \in \mathcal{V}} \left(y_i - y_i^{pred}\right)^2}$.
- MAE/MAD: the mean absolute error/deviance given by $\frac{1}{n_v} \sum_{i \in \mathcal{V}} \left|y_i - y_i^{pred}\right|$.
- R^2: just a transformation of RMSE, expressed as a percentage of improvement in mean square error (MSE) in comparison to the constant model. It can be calculated as $R^2 = 1 - \text{MSE}/\text{MSE}_0$ where $\text{MSE} = \text{RMSE}^2$, $\text{MSE}_0 = \frac{n_v - 1}{n_v} S_V^2$ and S_V^2 is the variance of Y in the validation dataset.
- (Negative) Log predictive score or density (LPS): is given by $-\log f(\boldsymbol{y}^{\mathcal{V}} | \boldsymbol{\theta}^{\mathcal{T}})$; where $\boldsymbol{y}^{\mathcal{V}} = \{y_i : i \in \mathcal{V}\}$ and $\boldsymbol{\theta}^{\mathcal{T}}$ are the parameters of the model estimated with the data of the train test.

For binary categorical factors the most common approach is to construct the 2×2 contingency table between the predicted and observed categories which is called confusion matrix in statistics and machine learning domains. The confusion matrix has the form of Table 2.7, where n_{ij} (for $i = 1, 2$ and $j = 1, 2$) are the frequencies for the combination of true and predicted outcomes (the values of one and two correspond to the negative and positive outcome respectively while i is used to the true outcome and j for the predicted one). The marginal frequencies $n_{i\bullet}$ and $n_{\bullet j}$ correspond to the marginal frequencies of the true and predicted outcomes, respectively.

It is obvious that the frequencies in the diagonal of the table correspond to the correct predictions and are called true negatives (TN=n_{11}) and true positives (TP=n_{22}), respectively. These observations are also called concordant

TABLE 2.7
Confusion matrix, predicted vs observed results

Actual Outcome	Predicted Outcome: Negative (N)	Predicted Outcome: Positive (P)	Marginal for True outcome
Negative (N)	n_{11} (TN)	n_{12} (FP)	$n_{1\bullet}$
Positive (P)	n_{21} (FN)	n_{22} (TP)	$n_{2\bullet}$
Marginal for Predicted	$n_{\bullet 1}$	$n_{\bullet 2}$	n

pairs (of predictions and observed values). Naturally, the off-diagonal frequencies n_{12} and n_{21} correspond to discordant pairs or values and represent the observations that are not predicted correctly. Specifically these are called false negatives (FN=n_{21}) and false positives (FP=n_{12}).

From these values, we calculate and report the following measures:

- *Sensitivity* or *recall* is the proportion of positive cases predicted correctly, given by $TP/n_{1\bullet}$.
- *Specificity* is the proportion of negative cases predicted correctly, given by $TN/n_{2\bullet}$.
- *Accuracy* is simply the proportion of correct predictions, given by $(TP + TN)/n$.
- *Cohen's Kappa* coefficient: this metric adjusts for correct prediction found by chance and is given by

$$\kappa = \frac{Accuracy - P_e}{1 - P_e} = 1 - \frac{1 - Accuracy}{1 - P_e}, \text{ where } \; p_e = \frac{1}{n}\sum_{j=1}^{J} n_{j\bullet} n_{\bullet j},$$

  and J is the number of levels for the outcome variable (in our case $J = 2$).

For general multi-category outcomes, we may use the accuracy, the kappa of Cohen and its weighted version. More details about evaluation metrics can be found in Section 3.4.

2.10 Summary and closing remarks of Chapter 2

In this chapter, we provided a comprehensive examination of statistical methodologies and computational tools applied in modelling football match outcomes. The discussion began with the formulation of predictive and descriptive models, particularly emphasizing the double Poisson model and its foundational role in football analytics. This model's simplicity, adaptability, and relevance to the sport's unique characteristics make it an essential starting point for further statistical exploration.

Key methods of estimation were reviewed, contrasting the classical approach of maximum likelihood estimation (MLE) with Bayesian inference. The chapter underscored the flexibility and depth offered by Bayesian methodologies, supported by tools like Markov Chain Monte Carlo (MCMC) algorithms, including Metropolis-Hastings and Gibbs sampling. These methods enable the integration of prior knowledge, allowing for a richer and more robust model calibration.

The inclusion of examples, such as the English Premier League dataset analysis, illustrated the practical implementation of these models. The use of R-based tools, including glm for MLE and advanced packages for Bayesian computation, provided actionable guidance for practitioners. The chapter also highlighted specialized Bayesian analysis software like WinBUGS/OpenBUGS which support advanced Bayesian modelling, and emphasized the critical role of careful data preparation and transformation.

Beyond methodology, the discussion addressed fundamental challenges and considerations in model application, including assumptions of independence, over-dispersion, and the dynamic nature of team abilities. The limitations of vanilla Poisson models, particularly in hybrid tournament formats, were acknowledged, advocating for the inclusion of team-specific covariates and performance metrics for enhanced prediction accuracy.

While this chapter has laid a solid foundation for understanding the statistical modelling of football outcomes, several avenues for future research remain. The integration of more sophisticated covariates, such as economic indicators or advanced performance metrics, could provide deeper insights. Additionally, exploring the scalability of these models to other team sports or non-sporting contexts could further validate their utility.

The emergence of computationally intensive methods, such as Hamiltonian Monte Carlo, signals a growing intersection of statistical rigor and computational advancements. Future efforts should focus on simplifying these tools' accessibility, fostering broader adoption among analysts and researchers.

In summary, this chapter has not only outlined the theoretical and computational frameworks necessary for football analytics but also set the stage for innovative applications in predictive modelling and decision-making within and beyond sports.

In the next chapter, we will focus in prediction for tournaments based on simulation. Detailed approaches based both on Bayesian and classical approaches will be presented in detail.

Appendix: Notation

Indexes and basic constants

- n: Number of games in the league/dataset
- K: Number of teams in the league/dataset
- T: Number of weeks in the league/dataset
- J: Number of levels/categories in a categorical variable; $J = 3$ for football match outcome.
- $i \in \{1, \ldots, n\}$: Observation/game index for game-arranged data
- $\ell \in \{1, 2\}$: Index denoting the home or away team for values one or two, respectively (for game-arranged data)
- $k \in \{1, \ldots, K\}$: Team index
- $w \in \{1, \ldots, W\}$: Week index
- $\imath$: observation index for univariate-arranged data with $\imath = 2i - 2 - \ell$
- $t \in \{1, \ldots, T\}$: Monte Carlo iteration index/superscript
- $j \in \{1, \ldots, J\}$: level/category index for a categorical variable

Model parameters

- $\boldsymbol{\theta}_{i1}, \boldsymbol{\theta}_{i2}$: parameter vectors for home and away teams
- ρ_i: dependence parameter between home and away goals
- μ: constant parameter in the vanilla model
- $home$: home effect parameter
- att_k, def_k: Fixed attacking and defensive parameter of k team
- $att_{k,t}$, $def_{k,t}$: Dynamic/random attacking and defensive parameter of k team at week t
- σ_a^2 and σ_d^2: random abilities variances
- $\beta_j^{(\ell)}$: effect of covariate j of home ($\ell = 1$) or away team ($\ell = 2$)

Variables and data for game-arranged data

- Y_{i1}, Y_{i2}: goals of the home and away team for game i
- $Z_i = Y_{i1} - Y_{i2}$: goal difference for game i
- Ω_i: Outcome of i game with three possible values: 1:home win, 2:draw, 3:home loss; $\Omega_i = \mathcal{I}(Z_i > 0) + 2\mathcal{I}(Z_i = 0) + 3\mathcal{I}(Z_i < 0)$.
- $O_i = 2 - \Omega_i$: Outcome of i game with three possible values: –1: away win, 0: draw, 1:home win.
- $\lambda_{i\ell}$: Expected goals (in Poisson) of the home and away team for game i
- $\eta_{i\ell}$: Predictor of the home and away team for game i
- $X_{ij}^{(\ell)}$, $x_{ij}^{(\ell)}$: Covariates/Features for the home or away team

Variables and data for univariate-arranged data

- $Y_\imath$: goals scored in $\imath$ observation of +++ data; Note that $\imath = 2i - 2 - \ell$ hence $Y_\imath$ refers to the goals scored by the home team ($\ell = 1$) or the away team ($\ell = 2$) in game i
- $HT_\imath$, $AT_\imath$: covariates denoting the home and away teams
- $Home_\imath$: dummy variable denoting if the goals $Y_\imath$ were scored by a home team
- $X_{\imath j}^{(\ell)}$: Covariates/Features for the scoring team ($\ell = 1$) of $Y_\imath$ or the opponent team ($\ell = 2$) which receives the goals $Y_\imath$
- $Att_\imath$: attacking team which scores $Y_\imath$ goals
- $Def_\imath$: defending team which receives $Y_\imath$ goals

3

Tournament and game prediction via simulation

3.1 Game score and outcome prediction

In this chapter, we will focus on the different approaches of making predictions using our statistical models. We first present how we can estimate individual games. We then proceed on presenting computationally intensive methods on how we can regenerate a league or a tournament using the data of a full season. This can be used to evaluate the goodness-of-fit of the model. Using several alternative Monte Carlo approaches, we can generate data for the remaining games and use this for prediction and/or for evaluating the out-of-sample predictive ability of the model under study.

3.1.1 Final score prediction using point estimates

The most popular approach is to estimate the model parameters and then, based on these point estimates, to calculate the probabilities of each outcome (home win/draw/away win) and consider as prediction the outcome with the highest estimated probability. Similarly, we can use the estimates of the estimated expected goals (or rounded versions of them) as the predicted scores.

For these estimates, the maximum likelihood estimates (MLEs) of the model parameters $\widehat{\boldsymbol{\beta}}$ are commonly considered point estimates. Such MLEs are directly available by our GLM implementation functions. Then, for each game i, the expected number of goals $\widehat{\lambda}_{i1}$ and $\widehat{\lambda}_{i2}$ are calculated as simple functions of the MLEs (if a goal-based model is considered). Finally, the probabilities of each match outcome $\widehat{p}_i^{Home}, \widehat{p}_i^{Draw}, \widehat{p}_i^{Away}$ will be calculated by the assumed model distribution using the estimated expected goals (and other additional parameters, if needed). In the case we use outcome-based predictive models, such as the multinomial logistic regression, then the final outcome probabilities $\widehat{p}_i^{Home}, \widehat{p}_i^{Draw}, \widehat{p}_i^{Away}$ are directly obtained as functions of the MLEs of model parameters, $\widehat{\boldsymbol{\beta}}$.

Similarly, in the Bayesian setup we can follow exactly the same logic by using as point estimates the posterior mean or median values of the model parameters $\widetilde{\boldsymbol{\beta}}$. Then a point estimate of the final match outcome probabilities can be simply obtained as function of $\widetilde{\boldsymbol{\beta}}$. Although such strategy is the direct analogue of the approach used in the frequentist approach, it is not

DOI: 10.1201/9781003186496-3

recommended within the Bayesian approach. With the power of MCMC, it is feasible to obtain a sample directly from the posterior distribution of the posterior outcome probabilities. Thus the whole distribution is available (through the posterior sample) and hence considering directly the posterior mean or median of these probabilities is more preferable than considering functions of the posterior means/medians of the model parameters $\boldsymbol{\beta}$.

Finally, all these approaches focus on predicting the final outcome as the point estimation of the match outcomes. Even if we consider the sampling distribution of $\boldsymbol{\beta}$ or their posterior distribution within the Bayesian paradigm, they do not account for the additional variability which is assumed/inherited via the model itself.

In the following, we are going to present three simulation-based alternatives which estimate the probability of each match outcome by taking into consideration also the uncertainty introduced by the assumed model/distribution. This approach can be used to re-generate whole leagues or competitions and, by this way, to estimate also other quantities of interest such as the final ranking in the league or the expected league points of each team.

3.1.2 Plug-in Monte Carlo method

The simplest approach to re-generate the league and estimate the probability of each outcome is to use the point estimates of the model parameters discussed in the previous section. Without loss of generality, let us denote the point estimates by $\widehat{\boldsymbol{\beta}}$, the general procedure can be described by the steps proposed by algorithm 2.

Algorithm 2 Game Score Generation Using Plug-in Monte Carlo Method

---

Inputs: $\boldsymbol{\beta}$: model parameters
$\widehat{\boldsymbol{\beta}}$: Estimates of model parameters
Y_{ik}: Goals scored by the home ($k = 1$) and away ($k = 2$) team in game i
$f(\boldsymbol{\theta}_{ik})$: Assumed goal distribution in game i for home ($k = 1$) and away ($k = 2$) team
$\boldsymbol{\theta}_{i1}$ and $\boldsymbol{\theta}_{i2}$: model parameters of $\mathcal{D}$ for home and away teams
$\widehat{\boldsymbol{\theta}}_{i1}$ and $\widehat{\boldsymbol{\theta}}_{i2}$: Estimates of model parameters of $\mathcal{D}$ for home and away teams

For $t = 1, \ldots, T$ REPEAT:

1. Calculate $\widehat{\boldsymbol{\theta}}_{i1}$ and $\widehat{\boldsymbol{\theta}}_{i2}$ as functions of the estimated model parameters $\widehat{\boldsymbol{\beta}}$
2. Generate $Y_{ik}^{*(t)}$ from $\mathcal{D}(\widehat{\boldsymbol{\theta}}_{ik})$ for $k = 1, 2$ (home and away teams) and game i
3. Calculate the predicted/generated goal difference for game i from $Z_i^{*(t)} = Y_{i1}^{*(t)} - Y_{i2}^{*(t)}$
4. Calculate the predicted/generated match outcome of game i from $O_i^{*(t)} = \mathcal{I}(Z_i^{*(t)} > 0) - \mathcal{I}(Z_i^{*(t)} < 0)$

** Variable $O_i^{*(t)}$ takes values in $\{-1, 0, 1\}$ with $1 \rightarrow$ home win, $0 \rightarrow$ draw, and $-1 \rightarrow$ win of the away team; $\mathcal{I}(A)$ is an indicator variable which takes the value one if A is true or takes the value of zero, otherwise.*

In this procedure, T denotes the number of generated repetitions for each game i in the league/competition under consideration. At the end of the above algorithm, we will end up with a set of T samples for each game with predicted goals $Y_{i1}^{*(t)}$ and $Y_{i2}^{*(t)}$ scored by home and away teams, the goal differences $Z_i^{*(t)}$, and the final predicted outcomes $O_i^{*(t)}$. A simple frequency tabulation of $Z_i^{*(t)}$ and $O_i^{*(t)}$ will provide estimated probabilities about the score difference and the final outcome. Estimated probabilities and predictions for specific scores can be obtained by the cross-tabulation of $Y_{i1}^{*(t)}$ and $Y_{i2}^{*(t)}$.

3.1.3 Prediction using Bootstrap

Parametric Bootstrap approach

The plug-in approach in the previous section ignores the estimation error and the sampling variability of the estimated parameters $\boldsymbol{\beta}$ which should be taken into consideration. A very simple approach is to incorporate this variability in the algorithms described in Section 3.1.2 by generating random values from the sampling distribution of $\widehat{\boldsymbol{\beta}}$. So instead of using directly MLE values $\widehat{\boldsymbol{\beta}}$ in the above simulation-based approach for each drop generated replication $t = 1, \ldots, T$, we first generate a set of coefficients

$$\boldsymbol{\beta}^{(t)} \sim N(\widehat{\boldsymbol{\beta}}^{(t)}, \widehat{\Sigma}),$$

where $\widehat{\Sigma}$ is the (approximate) variance/covariance matrix calculated by the inverse observed information matrix. So, under this approach, algorithm 2 will be now replaced by algorithm 3.

Algorithm 3 Game Score Generation Using Parametric Bootstrap

STEP 1 : Calculate $\widehat{\Sigma} = \left[\mathcal{I}(\widehat{\boldsymbol{\beta}})\right]^{-1}$ with $\mathcal{I}(\widehat{\boldsymbol{\beta}})$ being a matrix of dimension $p \times p$ with elements $I_{ij} = \frac{\partial^2 \ell(\boldsymbol{\beta})}{\partial \beta_i \partial \beta_j}$; where $\ell(\boldsymbol{\beta})$ is the model log-likelihood given by $\ell(\boldsymbol{\beta}) = \sum_{i=1}^{n} \log f(\boldsymbol{y}_{i1}, \boldsymbol{y}_{i2}|\boldsymbol{\beta})$ and p is the length of $\boldsymbol{\beta}$

STEP 2 : Generate from $\boldsymbol{\beta}^{(t)} \sim N(\widehat{\boldsymbol{\beta}}^{(t)}, \widehat{\Sigma}^{-1})$

STEPS 3–5 : Set $\widehat{\boldsymbol{\beta}} = \boldsymbol{\beta}^{(t)}$ and follow steps 1–4 from algorithm 2 in Section 3.1.2.

A simplified approach is to ignore possible correlation between parameters and generate all coefficients from

$$\beta_j^{(t)} \sim N\left(\widehat{\beta}_j^{(t)}, se(\widehat{\beta}_j)^2\right).$$

So now the game score generation procedure will be simplified to the one described by algorithm 4.

Algorithm 4 Game Score Generation Using Simplified Parametric Bootstrap

---

STEP 1 : Calculate the standard errors of $\widehat{\beta}_j$ (for $j = 1, \ldots, p$) by the output of a model fit, or by using the formula $se(\widehat{\beta}_j) = \sqrt{\left[\frac{\partial^2 \ell}{\partial \beta_j^2}(\widehat{\boldsymbol{\beta}})\right]^{-1}}$ (approximate standard error based on the observed Fisher information)

STEP 2 : Generate from $\beta_j^{(t)} \sim N(\widehat{\beta}_j^{(t)}, se(\widehat{\beta}_j)^2)$

STEPS 3–5 : Set $\widehat{\boldsymbol{\beta}} = \boldsymbol{\beta}^{(t)}$ and follow steps 1–4 from algorithm 2 in Section 3.1.2.

In the case that the $se(\widehat{\beta}_j)$ is not available (even asymptotically), then we can use bootstrap to estimate it. This can be done by getting replications of samples (with replacement) from the actual sample, estimating in each sample the coefficient of interest and use the estimated standard deviation (across different samples) of the estimated values as the standard error. The same procedure can be also used for the variance/covariance matrix. Note that by obtaining a sample with replacement here we mean that we select which rows of the dataset will appear in each generated bootstrap sample. This approach is computationally more demanding since it requires to fit T models in order to obtain the estimated coefficients in each bootstrap sample. For this reason, a better alternative might be to use a full bootstrap approach to generate predictions and estimate the required predicted probabilities and scores. This procedure is described in the paragraph which follows.

Full Bootstrap approach

An alternative to the parametric bootstrap is to use the full bootstrap approach. Again here we generate T bootstrap datasets which will be comprised by considering randomly (with replacement) rows of the original dataset. For each bootstrap sample, we fit our model and obtain the corresponding coefficient and fitted values. Then, we generate predicted scores based on the model sampling distribution and we can calculate the final probabilities by the frequencies of each match outcome.

This approach is computationally more demanding than the simple parametric bootstrap since it requires to fit T models in order to obtain the estimated coefficients in each bootstrap sample. Moreover, it might require larger number of replications than the bootstrap approach for the estimation of each $se(\widehat{\beta}_j)$ in order to achieve similar levels of precision. So the full bootstrap approach can be summarized by algorithm 5.

Algorithm 5 Game Score Generation Using Full Bootstrap

STEP 1 : Take a full sample (with replacement) $S_1^{(t)}, \ldots, S_n^{(t)}$ from value from 1 to n, that is the vector $(1, 2, \ldots, n)$ – in R, use the syntax `sample(1:n,replace=T)`

STEP 2 : Select the observations (i.e. rows of data) which correspond to $S_i^{(t)}$ (for $i = 1, \ldots, n$) to obtain the bootstrap sample $y^{*(t)}$ and $X^{*(t)}$

STEP 3 : Fit the model with the bootstrap data $y^{*(t)}$ and $X^{*(t)}$ to obtain $\widehat{\beta}^{*(t)}$

STEPS 4–6 : Set $\widehat{\beta} = \widehat{\beta}^{*(t)}$ and follow steps 1–4 from algorithm 2 in Section 3.1.2

3.1.4 Bayesian prediction via MCMC

Under the Bayesian approach, we may generate match results and regenerate a sport competition by obtaining samples of $Y_{i1}^{*(t)}$ and $Y_{i2}^{*(t)}$, for $t = 1, \ldots, T$, from the (posterior) predictive distribution which is given by

$$f(Y_{i1}^*, Y_{i1}^* | \boldsymbol{y}_1, \boldsymbol{y}_2) = \int f(Y_{i1}^*, Y_{i1}^* | \boldsymbol{\beta}) f(\boldsymbol{\beta} | \boldsymbol{y}_1, \boldsymbol{y}_2) d\boldsymbol{\beta},$$

where $\boldsymbol{y}_1, \boldsymbol{y}_2$ are the observed goals for the two opponent teams, and Y_{i1}^*, Y_{i1}^* are the random variables representing predicted future responses.

Under this approach, we work under a two-step simulating procedure: we only need to generate a sample of the model parameters $\boldsymbol{\beta}^{(t)}$ from the posterior distribution $f(\boldsymbol{\beta} | \boldsymbol{y}_1, \boldsymbol{y}_2)$ (second part of the above integral) and then obtain the samples $Y_{i1}^{*(t)}$ and $Y_{i2}^{*(t)}$ from the model sampling distribution $f(Y_{i1}^*, Y_{i1}^* \mid \boldsymbol{\beta})$ (first part of the integral) for the sampled parameters values, i.e. for $\boldsymbol{\beta} = \boldsymbol{\beta}^{(t)}$.

A standard way to obtain a sample from the posterior distribution $f(\boldsymbol{\beta} | \boldsymbol{y}_1, \boldsymbol{y}_2)$ is by using Markov Chain Monte Carlo (see for example in Ntzoufras, 2009) and related softwares such as WinBUGS/OpenBUGS (Spiegelhalter et al., 2003; Ntzoufras, 2009; Lunn et al., 2013), JAGS (Plummer, 2003) or Stan (Carpenter et al., 2017). Then, generating values from the predictive distribution corresponds to just adding the following line

$$(Y_{i1}^{*(t)}, Y_{i2}^{*(t)}) \sim f(Y_{i1}^*, Y_{i1}^* \mid \boldsymbol{\beta}^{(t)})$$

in each MCMC iteration of the algorithm. Moreover, the procedure can be also implemented by using a simple for-loop (or equivalent) code syntax even when a sample $\boldsymbol{\beta}^{(t)}$ for $t = 1, \ldots, T$ of the parameter vector $\boldsymbol{\beta}$ is available from the posterior distribution either from the MCMC output or any other Monte Carlo method. The procedure is described in algorithm 6 and can be implemented either within each MCMC iteration/step or after we obtain the MCMC output of the model parameters $\boldsymbol{\beta}^{(t)}$ for $t = 1, \ldots, T$.

Algorithm 6 Game Score Generation Using MCMC

Inputs: $\boldsymbol{\beta}$: model parameters

$\boldsymbol{\beta}^{(t)}$: model parameters generated at iteration/step t of the algorithm

$f(\boldsymbol{\beta}|\boldsymbol{y})$: the posterior distribution of the model parameters $\boldsymbol{\beta}$

$\mathcal{D}(\widehat{\boldsymbol{\theta}}_{ik})$: Assumed goal distribution $\mathcal{D}(\widehat{\boldsymbol{\theta}}_{ik})$ for $k = 1, 2$ (home and away games) and game i

$\widehat{\boldsymbol{\theta}}_{i1}$ and $\widehat{\boldsymbol{\theta}}_{i2}$: Estimates of model parameters of $\mathcal{D}$ for home and away games

For $t = 1, \ldots, T$ REPEAT:

1. Generate $\boldsymbol{\beta}^{(t)}$ from the posterior distribution $f(\boldsymbol{\beta}|\boldsymbol{y})$ using MCMC or other similar methods
2. Calculate $\boldsymbol{\theta}_{i1}^{(t)}$ and $\boldsymbol{\theta}_{i2}^{(t)}$ as functions of the generated values $\boldsymbol{\beta}^{(t)}$ of the model parameters $\boldsymbol{\beta}$ at t iteration
3. Generate $Y_{ik}^{*(t)}$ from $\mathcal{D}(\boldsymbol{\theta}_{ik}^{(t)})$ for $k = 1, 2$ (home and away games) and game i
4. Calculate the predicted/generated goal difference for game i from $Z_i^{*(t)} = Y_{i1}^{*(t)} - Y_{i2}^{*(t)}$
5. Calculate the predicted/generated match outcome of game i from $O_i^{*(t)} = \mathcal{I}(Z_i^{*(t)} > 0) - \mathcal{I}(Z_i^{*(t)} < 0)^*$

* *Variable* $O_i^{*(t)}$ *takes values in* $\{-1, 0, 1\}$ *with* $1 \rightarrow$ *home win,* $0 \rightarrow$ *draw, and* $-1 \rightarrow$ *win of the away team;*

$\mathcal{I}(A)$ *is an indicator variable which takes the value of one if* A *is true or takes the value of zero, otherwise*

3.2 Game outcome prediction from outcome-based models

In case we are using outcome-based models, then the procedure will be similar to the one described in algorithm 2 for the plug-in approach with the following changes:

1. The random variable of the outcome Y_i will take three values: 1, 2, 3 which correspond to the win, draw and loss of the home team, respectively.
2. Parameters $\boldsymbol{\theta}_i$ of each game i will be now the probabilities of win, draw and loss for the home team.
3. The outcome distribution $f(\boldsymbol{\theta}_i)$ will be now a multinomial distribution with three possible outcomes.

Hence, algorithm 2 will be now changed to algorithm 7 which follows.

Algorithm 7 Outcome Generation Using Plug-in Monte Carlo Method

Inputs: $\boldsymbol{\beta}$: model parameters

$\widehat{\boldsymbol{\beta}}$: (MLE) Estimates of model parameters

Ω_i: Outcome of game i with three possible values: 1 for home win, 2 for draw, 3 for away win

$O_i = 2 - \Omega_i$: Outcome of game i with three possible values: –1 for away win, 0 for draw, 1 for home win

$\mathcal{M}ultinomial(p_i^{Home}, p_i^{Draw}, p_i^{Away})$: Multinomial distribution for the outcome of game i with three possible outcomes: home win, draw, away win

$p_i^{Home}, p_i^{Draw}, p_i^{Away}$: model parameters of the $\mathcal{M}ultinomial$ distribution

$\widehat{p}_i^{Home}, \widehat{p}_i^{Draw}, \widehat{p}_i^{Away}$: Estimates of outcome probabilities.

For $t = 1, \ldots, T$ REPEAT:

1. Calculate the outcome probabilities $\widehat{p}_i^{Home}, \widehat{p}_i^{Draw}, \widehat{p}_i^{Away}$ for game i as simple functions of the estimated model parameters $\widehat{\boldsymbol{\beta}}$.
2. Generate the match outcome $\Omega_i^{*(t)}$ from the Multinomial distribution with probabilities $\widehat{\boldsymbol{p}}_i = \left(\widehat{p}_i^{Home}, \widehat{p}_i^{Draw}, \widehat{p}_i^{Away}\right)$.
3. Set $O_i^{*(t)} = 2 - \Omega_i^{*(t)}$ (in order to have the same coding for the match outcomes as in the algorithm for the goal-based approach).

The output of the algorithm will be now a vector of the possible predicted/generated outcomes $O_i^{*(t)}$ for each game i over T different repetitions. Again, the results can be summarized by the relative frequencies of the three outcomes over the T repetitions of the predicted/generated results. With this approach, obviously, no inference can be done for the final score or the score difference since the models focus only on predicting the final outcome and not the score of each game.

Similarly the bootstrap and Bayesian approach, Algorithms 3–5 and 6, respectively, will be slightly changed as described in the beginning of this section. Detailed description is omitted for brevity.

3.3 Tournament regeneration and prediction

The next step in our analysis is to simulate scenarios for checking the goodness of fit and the predictive ability of the fitted model (or more general machine learningalgorithm) with respect the final ranking of the tournament and the points collected at the end of the season in the case of a round-robin (league) competition.

3.3.1 League regeneration and prediction

Here we first focus on full leagues tournaments based on a round-robin system with a possibility for play-offs in some occasions. Other types of knock-out-based tournaments are discussed in Section 3.3.4 which follows. This can be implemented in two different perspectives:

(a) Retrospective approach (at the end of the season): re-simulate all the game results. By this we reproduce the whole league again using simulated results based on the data of the whole season (or more seasons if a more general model is used).

(b) Predictive approach (at any point in the middle of the season): re-simulate the remaining game results at a given point of the season. By this we predict what will be the final ranking of the league or the competition of the season.

The first approach is mainly used to see what will happen (according to the fitted model and the given estimated team performance) if the league was replicated many times. Hence, it can be used to compare the observed final league with the results obtained by the simulated leagues. This approach can be used to check the goodness of fit of the model in terms of final league reproducibility.

The second approach will act in a predictive fashion, since the first part of the observed games of the league will act as the training dataset while the rest of the games at the test dataset. Hence, with this approach we can assess the predictive ability of the fitted model.

In both approaches, the input will be the set of simulated outcomes $O_i^{*(t)}$ for $i \in \mathcal{V}$ and $t = 1, \ldots, T$ or alternatively the goal differences $Z_i^{*(t)}$ for each game i, where n is the number of games in the league and T denotes the number of simulated leagues.

The difference in the two approaches is the set of games $\mathcal{V}$ for which we will need to generate their results and the set of data we have used to train our model. In the first approach, we will fit/train our model using the data of the full season, i.e. $\mathcal{T} = \{1, 2, \ldots, n\}$ and we will generate results for all the games, that is $\mathcal{V} = \{1, 2, \ldots, n\}$, i.e. $\mathcal{T} = \mathcal{V}$. In the second approach we have data available up to game n^*, hence, we will fit/train our model using only the available data, i.e. $\mathcal{T} = \{1, 2, \ldots, n^*\}$ and we will generate results for the remaining games, that is $\mathcal{V} = \{n^* + 1, n^* + 2, \ldots, n\}$, hence $\mathcal{V} = \overline{\mathcal{T}} = \mathcal{D} \setminus \mathcal{T}$, where $\mathcal{D}$ is the data of the full season with n matches/games. Hence in the second approach, we set $O_i^{*(t)} = O_i^{(t)}$ or $Z_i^{*(t)} = Z_i^{(t)}$ (i.e. the observed outcomes or goal differences respectively) for $i = 1, 2, \ldots, n^*$.

3.3.2 Calculating expected points and other league metrics

Using the set of simulated outcomes $O_i^{*(t)}$ or goal differences $Z_i^{*(t)}$, then we can calculate the:

1. Expected number of points earned by each team.
2. Distribution of the points that could be potentially earned by each team.
3. Expected ranking and/or the ranking under the expected number of points.
4. Probabilities of ending up in each position of the league.

All the above measures can be used either graphically or using simple measures to assess how close is the final ranking under the model with the one finally observed. Given that you decide that the model performs sufficiently well, you can also assess simple working hypotheses like whether a team over-performed or under-performed in comparison to what was expected under the fitted model or whether a champion deserved to win the league.

In both approaches, we are interested to calculate the expected number of points for each team k, for $k = 1, \ldots, K$ denoted by xP_k. This will be simply derived as the sample mean of the points $P_k^{(t)}$ of each team calculated in each iteration t of the Monte Carlo algorithm. In the case of full regeneration of the league, then

$$P_k^{(t)} = HP_k^{(t)} + AP_k^{(t)},$$

where $HP_k^{(t)}$ and $AP_k^{(t)}$ are the predicted home and away points generated from the fitted model at iteration t of the Monte Carlo algorithm for team k, respectively. These quantities will be calculated as

$$\begin{aligned} HP_k^{(t)} &= \sum_{i=1}^{n} \mathcal{I}(ht_i = k)\left(3W_i^{(t)} + D_i^{(t)}\right) \\ AP_k^{(t)} &= \sum_{i=1}^{n} \mathcal{I}(at_i = k)\left(3L_i^{(t)} + D_i^{(t)}\right) \end{aligned}$$

where $\mathcal{I}(A)$ is an indicator function taking the value of one when A is true and zero otherwise, $W_i^{(t)}$, $D_i^{(t)}$ and $L_i^{(t)}$ are zero-one indicators for the observed win, draw and loss of the home team, respectively, in game i given by

$$W_i^{(t)} = \mathcal{I}(Z_i^{*(t)} > 0) = \mathcal{I}(O_i^{*(t)} = 1), \quad (3.1)$$

$$D_i^{(t)} = \mathcal{I}(Z_i^{*(t)} = 0) = \mathcal{I}(O_i^{*(t)} = 0), \quad (3.2)$$

$$L_i^{(t)} = \mathcal{I}(Z_i^{*(t)} < 0) = \mathcal{I}(O_i^{*(t)} = -1), \quad (3.3)$$

and $Z_i^{*(t)}$ are the generated observed goal differences for game i in iteration t of the Monte Carlo algorithm. If an outcome-based model is used, then we can only use the generated outcomes $O_i^{*(t)}$ to specify W_i, D_i and L_i game outcome indicators.

As we have already mentioned, in the second case, where only n^* data points/games are available, then the only change in the above approach will be to substitute $Z_i^{*(t)}$ by the observed goal differences Z_i or outcomes O_i for all the games $i \leq n^*$. Under this approach, the predicted points for iteration t can now be rewritten as

$$P_k^{(t)} = HP_{k,n^*} + AP_{k,n^*} + HP_{k,n^*}^{(t)} + AP_{k,n^*}^{(t)}, \tag{3.4}$$

where HP_{k,n^*} and AP_{k,n^*} are the observed points collected by team k up to game day or week w^* (i.e. up to available game $n^* = w^*K/2$) in home and away games, respectively; $HP_{k,n^*}^{*(t)}$ and $AP_{k,n^*}^{*(t)}$ are the predicted points generated from the fitted model at iteration t of the Monte Carlo algorithm for team k after game day or week w^* in home and away games, respectively. Therefore, the components of (3.4) will be calculated by the following formulas

$$\begin{aligned} HP_k &= \sum_{i=1}^{n^*} \mathcal{I}(ht_i = k)\,(3W_i + D_i) \;\; \& \;\; HP_{k,n^*}^{(t)} = \sum_{i=n^*+1}^{n} \mathcal{I}(ht_i = k)\left(3W_i^{(t)} + D_i^{(t)}\right), \\ AP_k &= \sum_{i=1}^{n^*} \mathcal{I}(at_i = k)\,(3L_i + D_i) \;\; \& \;\; AP_{k,n^*}^{(t)} = \sum_{i=n^*+1}^{n} \mathcal{I}(at_i = k)\left(3L_i^{(t)} + D_i^{(t)}\right). \end{aligned}$$

Under both the approaches, the expected number of points will be simply calculated as

$$\mathsf{xP}_k = \frac{1}{T}\sum_{t=1}^{T} P_k^{(t)}.$$

As the output of the simulation algorithm, we will obtain a sample $\mathsf{xP}_k^{(t)}$ of size T with the generated points achieved by team k for each of the T generated leagues/scenarios. Then we can easily obtain a predicted final table using the mean or median points. We can further use standard deviations or quantiles to quantify the accuracy of the replication or uncertainty of the final outcome and boxplots or histograms to depict of the distribution of the final points for each team; see Figures 3.1 and 3.2 and Table 3.1 for an example.

Similarly, for each league we can obtain the rankings (in descending order) of each team for each t-th generated league which is given by

$$R_k^{(t)} = 1 + \sum_{\ell=1}^{K} \mathcal{I}\left(\mathsf{xP}_k^{(t)} < \mathsf{xP}_\ell^{(t)}\right).$$

Again the above rankings can be used to obtain the mean or median rank or quantiles to summarize the most probable rankings of the team when the league is replicated many times; see, for example, Figures 3.3 and 3.4.

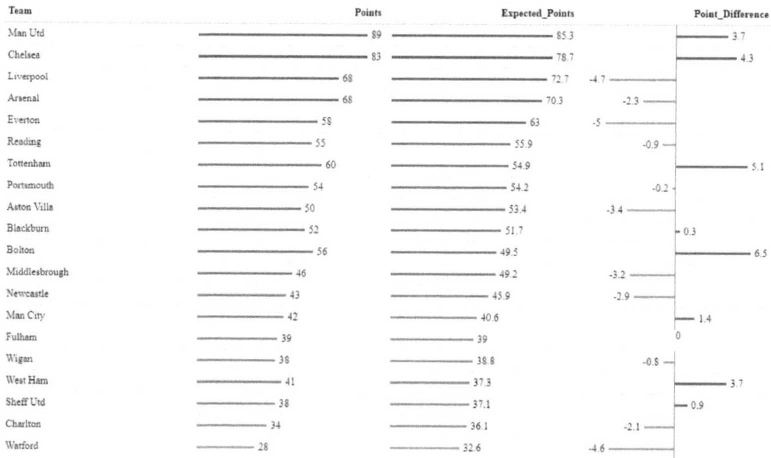

FIGURE 3.1
Comparison between observed and expected points for Premier League Data of Season 2006–2007.

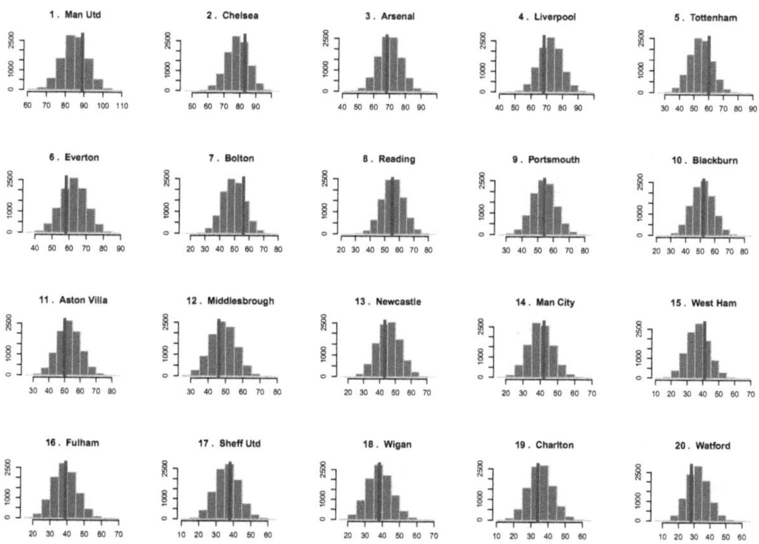

FIGURE 3.2
Distribution of points based on the Double Poisson model for Premier League Data of Season 2006–2007; the vertical reference line refers to the observed points.

TABLE 3.1
League Table based on simulated summaries for Premier League Data of Season 2006–2007

	Team	Expected Points	Stand. Deviation (Points)	95% CI Lower Bound (Points)	Median Points	95% CI Upper Bound (Points)	95% CI Lower Bound (Ranking)	Median Ranking	95% CI Upper Bound (Rankings)
1	Man Utd	85.3	6.7	72	85	98	1	1	4
2	Chelsea	78.7	7	65	79	92	1	2	5
3	Liverpool	72.7	7.2	58	73	86	1	3	7
4	Arsenal	70.3	7.4	56	70	85	1	4	8
5	Everton	63	7.4	49	63	78	2	5	11
6	Reading	55.9	7.6	41	56	71	4	8	15
7	Tottenham	54.9	7.5	40	55	69	4	8	15
8	Portsmouth	54.2	7.5	40	54	69	4	8	15
9	Aston Villa	53.4	7.5	39	53	68	4	9	16
10	Blackburn	51.7	7.4	37	52	66	4	10	17
11	Bolton	49.5	7.4	35	49	64	5	11	18
12	Middlesbrough	49.2	7.5	35	49	64	5	11	18
13	Newcastle	45.9	7.3	32	46	60	6	12	19
14	Man City	40.6	7.1	27	40	55	8	15	20
15	Fulham	39	7.2	25	39	54	8	16	20
16	Wigan	38.8	7.2	25	39	53	9	16	20
17	West Ham	37.3	7	24	37	51	9	17	20
18	Sheff Utd	37.1	6.9	24	37	51	10	17	20
19	Charlton	36.1	7	23	36	50	10	17	20
20	Watford	32.6	6.7	20	32	46	12	19	20

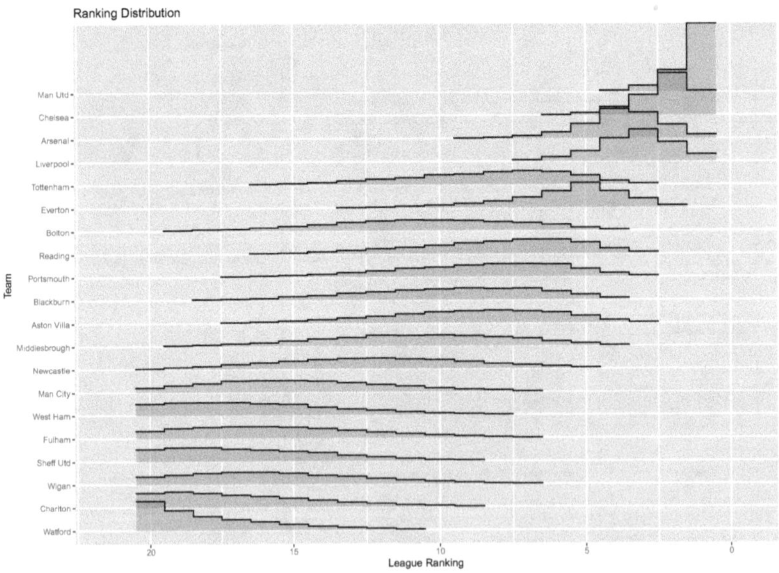

FIGURE 3.3
Distribution of league rankings based on the Double Poisson model for Premier League Data of Season 2006–2007.

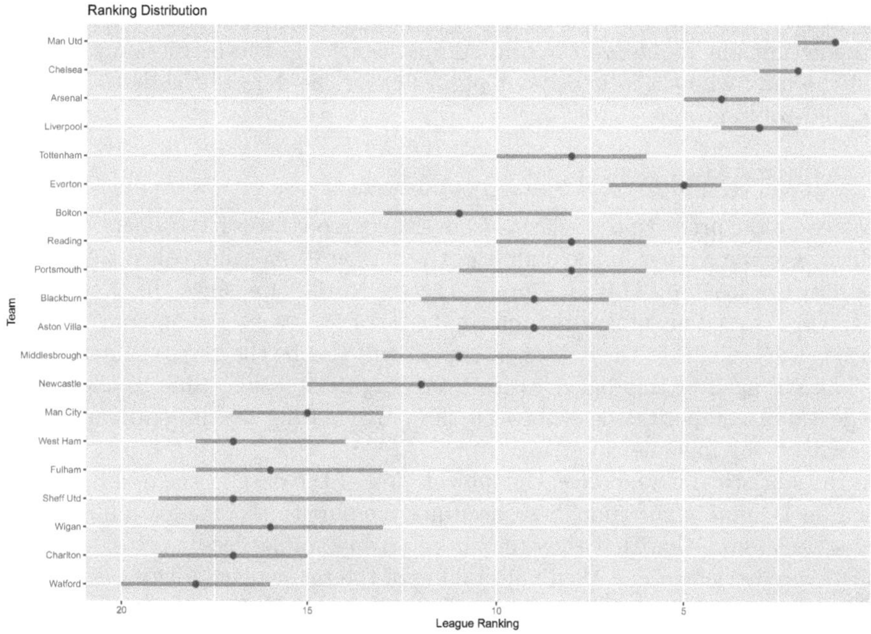

FIGURE 3.4
Error Bars based on quartiles of points obtained from the Double Poisson model for Premier League Data of Season 2006–2007.

3.3.3 League prediction scenarios

In order to evaluate the predictive ability of the league, we need to perform out-of-sample evaluations. The usual random split cross-validation techniques should not be implemented, especially if the model accounts for the temporality of the games. Hence, the most usual scenarios for evaluating the predictive ability in football (and possibly more general in sports) are the following

1. **Mid-season: prediction** use the data in the middle of the season to predict the final league table (usually in full round-robin competitions).
2. **One week-ahead: prediction** use the data up to a timepoint (week or matchday), in order to predict the games in the fixture of next week/matchday.
3. **Play-off: prediction** although Play-offs are not so common in football, when they exist it is very natural to use the full season data in order to predict the final winner in play-offs.
4. **Next-round: prediction** use data of the previous rounds to predict the next one (and the final winner). This is common in knock-out or hybrid[1] competitions.

[1]Hybrid competition is characterized from a first round with mini round-robin groups. The best teams (1 or more depending on the format) qualify in knock-out type of games.

5. **Knock-out Phase prediction** this is quite similar to the play-off prediction but it refers to hybrid competitions and the prediction of the final winners in knock-out phases using the data of the first round-robin groups phase.

3.3.3.1 Mid-Season prediction

The most common prediction scenario, from the perspective of the fans, is the prediction of the winner in the middle of the season (sometimes called the first round of the league). This is of great interest to the fans since the first team in the first round also takes the informal title of the "winter champion". There is also the belief that the winter champion will also be the league's final winner. The strength of this belief varies from league to league (due to the different levels of competitiveness between the participating teams) and can be extremely strong in some countries.

From the statistical perspective, the information of the data in the middle of the season is more than enough to accurately estimate the team's abilities (under the assumption that they remain relatively stable) leading to safe probabilistic predictions about the final ranking of the teams. Under this perspective, n^* is set equal to $K(K-1)/2$, $\mathcal{T} = \{1, 2, \ldots, K(K-1)/2\}$ and $\mathcal{V} = \{K(K-1)/2 + 1, \ldots, K(K-1)\}$. Then all the measures of interest are generated and calculated as described in Sections 3.1 and 3.3.2, respectively.

3.3.3.2 One week ahead prediction

In round-robin tournaments, one-week ahead prediction prediction is popular and generally straightforward to implement. In terms of prediction, this scenario assumes that the data up to a game day or week w^* are available (i.e. $n^* = w^*K/2$ games are available) and we are interested to predict the outcomes of game day or week $w^* + 1$ (i.e. K/2 additional games). Hence, the train dataset here will be $\mathcal{T} = \{1, 2, \ldots, w^*K/2\}$ and $\mathcal{V} = \{w^*K/2 + 1, \ldots, (w^* + 1)K/2\}$. Prediction can focus again on the game score itself, the outcomes (in terms of probabilities) of these $K/2$ games of $w^* + 1$ game day, or even in the prediction of points and rankings after this extra week. This procedure will be repeated after the results of each week are available and the model parameters will be updated with additional information for each week. Within this scenario, the use of the Bayesian approach is also of important value since it can be implemented sequentially by using as a prior for the data of week $w^* + 1$, the posterior distribution of the previous w^*. This can considerably speed up computations and, under this perspective, there is no need to re-run our model using all data but only using the new data.

Finally, in terms of out-of-sample evaluation of the predictive ability of the model, then the one-week ahead prediction can be used to measure the success of our adopted predictive model in each week (concerning goal score difference, accuracy, precision, recall or F1 of the game outcome, or in terms of points or ranking prediction) and then obtain the average or the distribution of the measure of interest across all seasons. One disadvantage of this approach is that we will need to re-run the model and the simulation after each week, and this can be alleviated by using Bayesian sequential methods.

3.3.4 Hybrid tournaments

In this section, we will focus on tournaments which is a combination of different tournaments: usually a first round-robin league followed by a knock-out phase or play-off phase. Here we will refer to some of such formats and then we will focus on prediction of the second phase after having observed the data of the first phase. We can categorize the different formats of hybrid tournaments in the following formats.

1. **Round Robin followed by Knock-out tournament**: at the end of the season, the top-K teams start a mini knock-out tournament where the two opponents play two times each other (once in each home stadium).
2. **Round Robin followed by Play-off Knock-out games with M_{req} number of wins**: at the end of the season, the top-K teams start a mini knock-out tournament where the two opponents play until one of the two teams wins N_{req} number of games. Usually the teams with the higher rank in the round-robin phase have priority in playing at their home stadium. This type of hybrid tournament is typical in other sports such as Basketball and Volleyball but not in Football.
3. **Multiple Round Robin Groups followed by Knock-out tournament**: the participating teams are separated in groups of K (usually four) teams according to their strength/ranking in previous seasons. The groups are composed from one team from each level of ranking in order that the competitiveness is similar across groups. The first and the second of each group proceed to the next knock-out phase. This is quite typical in Champions League, Europa League, and World Cups.
4. **Round Robin followed by Play-off Round-robin tournament**: not met very often but leagues with lower level of competitiveness (such as the Greek and the Scottish league) are using this format. Usually the top-K are separated in the second phase and the play again a full round-robin tournament. The points (or part of their points) in the first phase are carried along the second phase.
5. **Round Robin followed by Play-out (Knock-out) games**: the bottom-K teams play in a knock-out tournament with one or two games in each pair of opponents. Sometimes it is just a single knock-out phase deciding the winners.

There are also other special types of tournaments where their design is rather creative. Such tournaments can include different layers of different phases and types of tournaments (round robin, simple knock-out phase or play-off knock-out games with multiple games). For example, in the Greek Superleague currently comprises from 14 teams. The league has both play-off and play-outs in the form of round-robin tournaments. The play-out include the bottom eight teams which they compete in order to avoid the bottom two positions leading to relegation. In the same league the top six teams qualify for the play-offs (again in round-robin format) and they compete each other for the title of the champion as well as for the remaining tickets for the European tournaments (Champions League, Europa League and Conference League). In the past, the champion was announced in the normal season and the play-offs were between teams in positions 2–7 which were competing each other for places either in Champions league or in Europa league.

3.3.4.1 Knock-Out Play-off prediction

This format is not popular in association football but in sports like basketball or volleyball. We have included it in this section, in order to make this chapter more complete in terms of different tournament formats.

Usually, a knock-out play-off phase follows after the end of a regular round-robin season. Each pair of opponents play in a series of games. The winner is the team which first wins N_{req} games in the series. Hence, the minimum number of matches between two opponents is at least N_{req} and at most $N_{seq} = 2N_{req} - 1$ with N_{req} to be usually equal to three, five or seven (usually an odd number of games). Often N_{req} increases as the tournament progresses. For example, we may have $N_{req} = 3$ for the quarter-finals, $N_{req} = 5$ for the semi-finals and $N_{req} = 7$ for the final. The games are in the home stadiums of both teams in a scheme which interchanges. Usually, a team has an advantage and plays the first and the last game of the series in their home stadium.

For each simulation scheme, we generate a sequence of matches from the predictive distribution until we announce a qualifying team; see algorithm 8 for a detailed description. An alternative, simpler way to identify the winner in each pair of opponent teams in algorithm 8 is to always generate $N_{seq} = 2N_{req} - 1$ matches and then announce as the qualifying team the one with the highest number of wins. Although, this approach is equivalent to the algorithm 8 in terms of results, it is less efficient since we are required to generate a larger number of matches than the ones needed in order to announce the winner.

For such phase, we may focus on reporting the percentage of correct predictions with respect to the winner of each game or the team which qualified to the next round followed by an analysis about the final score or the game goal difference in each game. Although in other sports N_{req} is equal to 3 or 5, in football usually we have only two games (one in each home stadium) and it is usually referred simply as a knock-out phase. Also the final game in such knock-out phases or tournaments consist of a single game in a neutral stadium (i.e. $N_{req} = 1$).

Algorithm 8 Stochastic play-offs prediction algorithm

For $t = 1, \ldots, T$ REPEAT:

Set $i = 0$

Set $g = 0$, $W_1 = 0$ and $W_2 = 0$

While $W_1 < N_{req}$ **or** $W_2 < N_{req}$

Update $g = g + 1$ and $i = i + 1$

Update which team plays at home (ht) and which as away team (at) in the g match in the series of games

Update the parameters of the model accordingly

Generate the match outcome $\Omega_i^{*(t)}$ from the assumed model (Poisson, multinomial or other)

Set $O_i^{*(t)} = 2 - \Omega_i^{*(t)}$ (-1 indicates that the away wins, 0 a draw and 1 win of the home team)

If $O_i^{*(t)} = 1$ **then** $W_1 = W_1 + 1$ **else If** $O_i^{*(t)} = -1$ **then** $W_2 = W_2 + 1$

If $W_1 = N_{req}$ **then** $Q^{(t)} = 1$ **else if** $W_2 = N_{req}$ **then** $Q^{(t)} = 2$

Return Q

Indexes:
$t =, 1, \ldots T$; T: *number of simulated values for each game;*
$k = 1, \ldots, N_{pairs}$; N_{pairs}: *number of pairs of opponents to be predicted; each pair of opponents corresponds to a sequence of matches until* N_{req} *wins are reached from one of the opponent teams.*
W_1, W_2: *number of wins for home and away teams of pair* k*, respectively;*

Simulation of Knock-out play-off phase or tournament is described in algorithm 8 with the simple knock-out phase or tournament to be a special case with $N_{req} = 2$ for all series of games except for the final game where $N_{req} = 1$ as discussed in the previous paragraph.

3.3.4.2 Knock-out prediction

The simple knock-out phase or tournaments which is a frequent scheme in European competitions (Champions League knock-out phase, National cup tournaments etc.) are easier to be simulated since the number of required games is constant and equal to $N_{req} = 2$ (playing in home and away stadium of a team) or $N_{req} = 1$ (in finals or European or World Cups of National teams). Such a tournament can be implemented using 8 with $N_{seq} = 2$ or N_{seq} and a slight modification on announcing the winner when we have an overall tie between the competitors; see modified algorithm 9 for details.

Algorithm 9 Stochastic knock-out prediction algorithm

For $t = 1, \ldots, T$ REPEAT:

Set $W_1 = 0$

For $i \in \{1, 2\}$

Update which team plays at home (ht) and which as away team (at) in the g match in the series of games

Update the parameters of the model accordingly

Generate the match outcome $\Omega_i^{*(t)}$ from the assumed model (Poisson, multinomial or other)

Set $O_i^{*(t)} = 2 - \Omega_i^{*(t)}$ (-1 indicates that the away wins, 0 a draw and 1 win of the home team)

$W_1 = W_1 + O_i^{*(t)}$

If $W_1 > 0$ **then** $Q^{(t)} = 1$
else if $W_1 < 0$ **then** $Q^{(t)} = -1$
else Consider the goal difference or penalties to announce the winner

Return Q

Indexes:
$t =, 1, \ldots T$; T: *number of simulated values for each game.*

Prediction of the second phase in other hybrid tournaments

The final output of the Monte Carlo of MCMC algorithms described in Sections 3.1–3.2 is a matrix of values of the final score or the final outcome (in the form of home win/draw/away win). Although, some metrics on the game level can be implemented, it is difficult to find a model which will have an increased precision. Hence, we focus on testing the fit and the predictability with respect of some marginal characteristics such as the overall distribution of game scores, goal differences or the accuracy in the reproduction of the final league.

3.4 Measures of goodness of fit and predictive performance

When implementing a predictive model, it is desirable to evaluate its performance and, where appropriate, compare it with other competing models or methods. This evaluation is typically conducted using selected measures that compare the actual outcomes with those predicted by the model.

There are two different but complementary perspectives for this evaluation. The first, known in statistical modelling as "goodness of fit", involves using measures to compare the actual responses with the predicted or fitted values for all observations that were used to estimate the model parameters (i.e., the data used for learning). This process is referred to as "in-sample" model evaluation. Ideally, we want a model that is well-fitted, meaning it explains or predicts the data used for learning in a satisfactory manner. A poorly fitted model will also produce inaccurate future predictions.

However, having a well-fitted model does not ensures a good predictive performance. Goodness-of-fit measures can often overestimate a model's predictive ability, and in the worst case, you may end up with a model that fits the current data perfectly but leaves no room for uncertainty in predicting future observations. Such models are called to be over-fitted. As a result, the goodness-of-fit approach only helps us eliminate models with very poor predictive performance.

Therefore, in order to properly evaluate our models, we also need to use "out-of-sample" approaches, which involve calculating prediction measures based on responses (i.e. match outcomes) that were not used in the model's estimation or learning process. This approach provides a more realistic assessment of the model's predictive ability. Ultimately, the goal is to identify models that exhibit both acceptable goodness-of-fit and the best out-of-sample predictive performance.

In this section, we discuss measures that can be used for both "goodness-of-fit" and "predictive" evaluations. The key distinction between these two approaches lies in the data used for their calculation. When we use the training dataset (i.e., the same data used for estimation or learning), we obtain a goodness-of-fit measure. On the other hand, when we use a test dataset (i.e., observations not involved in the estimation or learning process), we derive a predictive evaluation measure.

This review focuses on predictive measures based on score outcomes, as well as purely probabilistic performance measures, known as *scoring rules.* The latter are based on the difference between the probabilities assigned (or derived) for specific events and the actual outcomes.

3.4.1 Root mean absolute error and mean absolute error

The most commonly used measure for comparing observed and predicted numerical outcomes is the Root Mean Squared Error (RMSE). As the name suggests, RMSE calculates the square root of the average squared differences between observed and predicted (or expected) values. In football, RMSE can be applied to assess the accuracy of predictions in several ways. At the match level, it can evaluate the difference between actual and predicted goals scored or the predicted goal difference. Additionally, RMSE can be used to measure the differences between observed points (in the final league standings) and predicted points (obtained by re-generated leagues as described in Section 3.3) for each team. Hence, for observed outcomes y_i and predicted values $\widehat{y}_i$, for $i = 1, \ldots, n$, the RMSE is given by

$$\text{RMSE} = \sqrt{\frac{1}{n}\sum_{i=1}^{n}(y_i - \hat{y}_i)^2}, \tag{3.5}$$

where n is the sample size of the dataset used for evaluation (i.e., the size of either the training or test dataset). Have in mind, that when we refer to regression models and we apply the formula 3.5 on training data, then the RMSE is simply an (biased) estimate of the regression error standard deviance (or the square root of the residual sum of squares).

A key property of RMSE is its sensitivity to large differences between actual and predicted outcomes. This characteristic can be useful for identifying and avoiding large deviations in predictions. However, it also means that one or a few poor predictions can deteriorate the overall evaluation metric, potentially failing to correctly record the model's overall predictive performance.

Alternatively, instead of considering we mean of the squared differences and the take their square root, we can directly consider the mean of the absolute differences and obtain the Mean Absolute Error (MAE) given by

$$\text{MAE} = \frac{1}{n}\sum_{i=1}^{n}|y_i - \hat{y}_i|.$$

MAE avoids the need to square the differences and then apply the square root, as it directly considers absolute differences. Consequently, MAE is easier and more straightforward to interpret than RMSE because it presents errors in the same units as the response data. Furthermore, MAE is less sensitive to extreme individual prediction errors than RMSE. It is sometimes referred to as the Mean Absolute Difference or Deviance (MAD). In practice, RMSE and MAE are used in complementary way, and it is common to report both when comparing or evaluating different predictive models and methods.

With respect to the quantitative characteristics of football, we can calculate both RMSE and MAE for different quantitative measures either obtained from individual games or other marginal game outcomes such as the final league standings. In general we write

$$RMSE(Q^{(t)}) = \sqrt{\frac{1}{|Q|}\sum_{k=1}^{|Q|}\left(Q_k^{pred(t)} - Q_k\right)^2} \tag{3.6}$$

$$MAE(Q^{(t)}) = \frac{1}{|Q|}\sum_{k=1}^{|Q|}\left|Q_k^{pred(t)} - Q_k\right| \tag{3.7}$$

for $k = 1, \ldots, |Q|$ and $t = 1, \ldots, T$ where $|Q|$ is the length of vector Q on which the corresponding measure is based and T the number of generated values of Q. The above versions of RMSE and MAE consider also a time-point t in order to denote that these quantities can be also calculated in each iteration of a Monte Carlo method (or bootstrap) instead just comparing the point predictions of the model.

We may consider different quantities as Q in the calculation of predictive measures. In match level we may consider:

- the number of goals scored by each opponent ($Q_k = Y_{k1}$ and/or $Q_k = Y_{k2}$, for $k = 1, \ldots, n$; Y_{k1} and Y_{k2} are the goals scored by the home and the away team, respectively, in k game),
- the goal difference in each game ($Q_k = Z_k = Y_{k1} - Y_{k2}$, for $k = 1, \ldots, n$; where Z_k is the goal difference for game k),

while in the final league level, we can compare

- the number of points collected by each team (i.e. $Q_k = P_k$ for $k = 1, \ldots, N$; where P_k are the total points collected by team k and N is number of teams in the league),
- the total goal scored by each team (i.e. $Q_k = TG_{k1}$ for $k = 1, \ldots, N$; where TG_{k1} are the total goals scored by team k),
- the total goal conceded by each team (i.e. $Q_k = TG_{k2}$ for $k = 1, \ldots, N$; where CG_{k2} are the total conceded goals by team k),
- the total goal difference of each team (i.e. $Q_k = TZ_k$ for $k = 1, \ldots, N$; where $TZ_k = TG_{k1} - TG_{k2}$ is the total goal difference for team k)
- the ranking of each team (i.e. $Q_k = R_k$ for $k = 1, \ldots, N$; where R_k is the final ranking of team k).

Finally, we can assess the marginal distributions of goals scored by the home team, the away team, and the goal differences by comparing the observed and predicted frequencies (i.e., the number of matches) for each value of these three response variables.

3.4.2 Coefficient of determination

Coefficient of determination or simply R^2 is a measure that is primarily used in regression to evaluate the goodness-of-fit of a model and is given by

$$R^2 = 1 - \frac{\sum_{i=1}^{n} (y_i - \hat{y}_i)^2}{\sum_{i=1}^{n} (y_i - \overline{y})^2} = 1 - \frac{\widehat{\sigma}_\epsilon^2}{\widehat{\sigma}_y^2}$$

where $\widehat{\sigma}_\epsilon^2$ and $\widehat{\sigma}_y^2$ are the biased estimators for the error variance and the variance of the response variable Y which are given by

$$\widehat{\sigma}_\epsilon^2 = \frac{1}{n}\sum_{i=1}^{n}(y_i - \hat{y}_i)^2 \text{ and } \widehat{\sigma}_y^2 = \frac{1}{n}\sum_{i=1}^{n}(y_i - \overline{y})^2. \tag{3.8}$$

Hence, R^2 can be viewed as a simple transformation of the error variance. Since the variance components in both the numerator and denominator of R^2 share the same measurement units, this metric is free of units. Its interpretation is quite appealing since it expresses the proportion of variance of the response variable that is explained by the predictive model.

When applied to a training dataset, R^2 ranges from zero to one. A value of zero indicates a poor model fit, equivalent to the simplest possible predictive model. This model is often called the constant model and uses the sample mean as a prediction for all observations. A value of one indicates a perfect fit, meaning the predicted values exactly match the observed ones. However, a perfect fit should raise suspicions, as it often results from obvious relationships (such as using the same variable for both the response and explanatory variables), recurrent patterns, or severe overfitting.

Although R^2 is commonly used as a measure of goodness-of-fit in regression models applied to training data, it can also be easily implemented for test datasets. As mentioned in Section 3.4.1, for test data, the standard deviation of the error, $\sigma_\epsilon = RMSE$.

$$R^2 = 1 - \frac{RMSE^2}{\widehat{\sigma}^2_{y^{test}}}$$

where $\widehat{\sigma}^2_{y^{test}}$ is simply the (biased) estimate of the variance obtained from the test dataset. Hence, the predictive measure of R^2 can be considered as a simple transformation of RMSE.

3.4.3 Brier score

The Brier score was firstly introduced by Brier et al. (1950) for weather forecasting, and used, among the others, by Spiegelhalter and Ng (2009) for football prediction. Essentially, the Brier score is nothing more than an MSE (i.e. squared RMSE) adopted for categorical data. In its simple version, when binary outcomes are assumed, then y_i is the observed outcome denoted by o_i and $\hat{y}_i$ is replaced by the corresponding probability of this outcome p_i.

Over the years, the Brier score has become one among the most well-known indicators for measuring the plausibility of some predictions expressed in terms of probabilities. Perhaps, the Brier score is a probabilistic score function designed to assess the accuracy of probabilistic prediction, being defined as the mean-squared error (MSE) of the forecasts. According to its original formulation proposed by Brier et al. (1950), the score measures the mean squared difference between the predicted probability assigned to the possible outcomes for item i and the actual outcome a_i, ranging from zero to two. Therefore, the lower the Brier score is for a set of predictions, the better the predictions are calibrated: a Brier equal to zero corresponds to a perfect prediction, whereas a Brier equal to two to a useless prediction that put 100% probability on an outcome that did not occur. For unidimensional predictions, it is strictly equivalent to the MSE as applied to predicted probabilities.

We start by defining the Brier score for binary events. Regarding football, we could be interested in evaluating events such as E_n: *"The home team wins against the away team in the i-th match"*. In this case the Brier score is defined as:

$$\text{BS} = \frac{1}{n}\sum_{i=1}^{n}(p_i - o_i)^2, \tag{3.9}$$

where p_i is the probability for the home win, o_i is the observed outcome, equal zero if the event does not happen, one otherwise. Note that the version of the BS reported in Equation (3.9) takes values from zero to one. As we have already mentioned, $BS = RMSE^2$ when considering $\widehat{y}_i = p_i$ and $y_i = o_i$.

As an example, consider the English Premier League 1994/1995 played on May, 14th 1995, as reported in Table 3.2: the third column reports some artificial individual probabilities for the home team win, whereas the fourth and the fifth column report the draw and the away win probabilities, respectively. The sixth and seventh column indicate the number of home and away goals, respectively. The eighth column reports the actual observed result. Teams denoted in bold correspond to home teams winning the matches ($a = 1$). To compute the BS for binary events such as *"home win/no home win"* we just need the third column of probabilities. The BS according to Equation (3.9) is

TABLE 3.2
English Premier League 1994/1995 probabilities and observed results for the 42nd match-day, played on May, 14th

home team	away team	p^H	p^D	p^L	hg	ag	obs
Chelsea	Arsenal	0.61	0.17	0.22	2	1	H
Coventry	Everton	0.45	0.25	0.30	0	0	D
Liverpool	Blackburn	0.52	0.25	0.23	2	1	H
Manchester City	**QPR**	0.45	0.18	0.37	2	3	A
Newcastle	Crystal Palace	0.56	0.23	0.21	3	2	H
Norwich	Aston Villa	0.39	0.30	0.31	1	1	D
Sheffield Wed	Ipswich Town	0.55	0.21	0.24	4	1	H
Southampton	Leicester	0.55	0.26	0.19	2	2	D
Tottenham	Leeds	0.61	0.18	0.21	1	1	D
West Ham	Manchester United	0.35	0.25	0.40	1	1	D
Wimbledon FC	Nottingham	0.35	0.22	0.43	2	2	D

then given by:

$$\text{BS} = (0.61 - 1)^2 + (0.45 - 0)^2 + \ldots + (0.37 - 0)^2 + (0.35 - 0)^2 = 0.205.$$

We can extend the BS specification from binary forecasts to multi-category forecasts by introducing the original BS formulation. In this case we should now evaluate even the draw and the away probabilities and compute the BS as follows:

$$\text{BS}^{(J)} = \frac{1}{n} \sum_{i=1}^{n} \sum_{j=1}^{J} (p_i^j - a_i^j)^2, \tag{3.10}$$

where J is the number of categories of the response, $j = 1, \ldots, J$, p_i^j is the forecast probability of the outcome j in the i-th match, and a_i^j is a dummy coding for the actual outcome in the i-th match, equals one if event j happened, zero otherwise.

The above equation for a trivariate outcome as in football then $J = 3$ and (3.11) will simplify to

$$\text{BS}^{(3)} = 1 + \frac{1}{n} \sum_{i=1}^{n} \sum_{j=1}^{3} (p_i^j)^2 - \frac{2}{n} \sum_{i=1}^{n} p_i^{o_i}, \tag{3.11}$$

where p_i^1, p_i^2 and p_i^3 are the probabilities for win, draw and loss of the home team in i match, respectively, and $p_i^{o_i}$ is the probability of the observed outcome in i match. Note that for binary variables $\text{BS}^{(2)} = 2\,\text{BS}$.

We can apply the Brier score in (3.11) for multi-class predictions of the Premier League results reported in Table 3.2. First of all, we need to compute the internal sum in Equation (3.11) for each of the eleven matches considered. For the first match, Chelsea vs Arsenal, we have $(0.51 - 1)^2 + (0.27 - 0)^2 + (0.22 - 0)^2$, whereas for Coventry vs Everton we have $(0.45 - 0)^2 + (0.25 - 1)^2 + (0.3 - 0)^2$, and so on for the remaining matches. The final BS is equal to 0.648 (or 0.324 if you divide it by two in order to be bounded at one).

In general, how could we assess the prediction accuracy through the Brier score? As remarked above, the lower the Brier score, the better is the model's predictive accuracy. We could use an intuitive benchmark in the following way: if we had used a *naive classifier* assigning some uniform probabilities—therefore considering each outcome with equal probability, 0.333—instead of our modelling/individual classifier generating the probabilities included in Table 3.2, the BS would have been 0.666 (or 0.333 if we consider the half of it in order to be bounded at one), lower, thus better, than that obtained through individual probabilities. The latter is 2.8% lower than the BS of a random/naive classified. Thus, a rough measure of accuracy is represented by the comparison between the probabilistic predictions obtained with the naive/random classifier.

3.4.4 Ranked probability score

The Ranked Probability Score (RPS) (Epstein, 1969; Winkler, 1969; Murphy, 1970) measures the difference between the cumulative distribution function (CDF) of the forecasted probabilities and the CDF of the observed outcomes. It measures the sum of squared differences in cumulative probability space for a multi-category probabilistic forecast: therefore, it is calculated by summing the squared differences between the forecasted cumulative probabilities and the observed cumulative probabilities for each possible outcome, across all possible outcomes.

For a generic event with K categories, the RPS is defined as:

$$\text{RPS} = \frac{1}{n}\sum_{i=1}^{n} \text{RPS}_i \text{ with } \text{RPS}_i = \frac{1}{J-1}\sum_{j=1}^{J}\left(\sum_{l=1}^{j} p^l - \sum_{l=1}^{j} a^l\right)^2, \qquad (3.12)$$

where p^j is the predicted probability in forecast category j, whereas a^j is the indicator (0=no, 1=yes) for the observation in category j. The RPS in Equation (3.12) ranges between 0 and 1, with a lower score indicating better forecasting performance. A score of 0 indicates perfect forecasting, where the predicted probabilities match the observed probabilities exactly. A score of 1 indicates a completely inaccurate forecast, where the predicted probabilities are completely different from the observed probabilities. In case of binary events, it is straightforward that the RPS is equivalent to the BS in Equation (3.9).

According to a single football match, the number of categories is $K = 3$, which means $K - 1 = 2$. We could then average the RPS for a single match (3.12) over the n matches of interest and obtain then:

$$\text{RPS} = \frac{1}{n}\sum_{i=1}^{n}\frac{1}{2}\sum_{j=1}^{J}\left(\sum_{l=1}^{j} p^l - \sum_{l=1}^{j} a^l\right)^2, \qquad (3.13)$$

where p^j is the predicted probability in forecast category j, whereas a^j is the indicator (0=no, 1=yes) for the observation in category j.

RPS has been used in the past for the evaluation of football prediction models. Constantinou and Fenton (2012b) argue that the RPS is the most suitable metric for evaluating probabilistic forecasts in football matches. Consequently, the RPS has emerged as one of the most widely adopted and popular scoring rules for this application (Wheatcroft, 2021). Schauberger et al. (2016) used RPS to tune their hyperparameters and achieve maximum predictive performance. Baboota and Kaur (2019b) used RPS to evaluate their methods and algorithms or even tune their hyperparameters and achieve maximum predictive performance.

On the other hand, Wheatcroft (2021) strongly criticized the use of RPS for evaluating football predictions. He claimed that the reasoning of Constantinou and Fenton (2012b) in favour of RPS was oversimplistic and the conclusion questionable. From his experiments, Wheatcroft (2021) reported that the ignorance score outperformed both the RPS and the Brier scores with the latter to be better than RPS.

The global RPS in (3.13) ranges from zero and one and maintains the same interpretation as previously remarked: by applying it to the English Premier League results reported in Table 3.2, we obtain RPS = 0.158 which is about 15.7% lower (better) than the naive/random classifier with RPS=0.187.

3.4.5 Average of correct probability

Another typical assessment for the probabilistic predictive accuracy is given by the average of the *correct* probabilities, namely the average of the probabilities assigned to the outcomes that actually occurred. Let us denote this vector of probabilities with $p_1^*, p_2^*, \ldots, p_n^*$, then the average over the n matches is given by:

$$\text{ACP} = \frac{1}{n}\sum_{i=1}^{n} p_i^*, \tag{3.14}$$

which will be as high as the predictive accuracy is large. Conversely for what happens with the mean-squared error-based measures of the Brier and the Ranked Probability scores, this metric approaches 1 when the predictive accuracy is perfect, and 0 when is completely wrong. In practice, the ACP will rarely approach the extremes 0 or 1: in general, an ACP value greater than 0.33 for the single match denotes that the model/subjective probabilities are better than those suggested by the naive classifier. With respect to the English Premier League example reported in the previous sections, the ACP is equal to 0.341, suggesting that the performance is slightly higher than the random assignment given by the naive classifier.

3.4.6 Pseudo-R^2

Another common way to assess the probabilistic predictive performance over a number of matches is the pseudo-R^2, defined as the geometric mean of the probabilities assigned to the actual result of each match played during the forecasting period (Dobson et al., 2001). By using again the vector of probabilities $p_1^*, p_2^*, \ldots, p_n^*$ introduced in the previous section, the pseudo-R^2 is given by:

$$\text{pseudo-R}^2 = (p_1^* p_2^* \cdots p_n^*)^{1/n}. \tag{3.15}$$

As a matter of interpretation, the psuedo-R^2 will be as high as the probabilities for the outcomes actually observed will be high, approaching 1 when the decision maker place a probability exactly equal to one on the actual observed results, and zero when he/she places a probability of zero.

In the English Premier League example, the pseudo-R^2 is equal to 0.315, denoting again that the decision maker probabilities are not better than the uniform probabilities offered by a naive/random classifier.

3.4.7 Measures for assessing predictive performance for binary outcomes

When we consider binary responses, one of the most widely used approaches in statistics and machine learning to evaluate the predictive performance of a model is by constructing the *confusion matrix* and extracting various performance metrics from this matrix. The confusion matrix is a 2×2 contingency table that displays the joint distribution of the actual versus predicted outcomes. It classifies the predictions into four categories: true positives (TP), true negatives (TN), false positives (FP), and false negatives (FN). This structure allows for the assessment of a model's performance via the calculation of several simple and comprehensive measures of predictive performance. The format of the confusion matrix is shown in Table 3.3.

TABLE 3.3
Typical structure of a confusion matrix

Actual	Predicted Outcome ($\hat{Y}$)	
Outcome (Y)	No ($\hat{Y}=0$)	Yes ($\hat{Y}=1$)
No ($Y=0$)	n_{11} (TN)	n_{12} (FP)
Yes ($Y=1$)	n_{21} (FN)	n_{22} (TP)

TN: True Negative; FP: False Positive; FN: False Negative; TP: True Positive

More specifically we calculate the key: accuracy, precision, recall (sensitivity), specificity and the F1 score which are defined as follows.

3.4.7.1 Accuracy

Accuracy measure quantifies the proportion of correct predictions of our model and mathematically is given by the formulae:

$$\begin{aligned} \text{Accuracy} &= \text{Pr(Correct Predictions)} = \frac{\text{Number of correct predictions}}{\text{Total number of observations}} \\ &= \frac{\text{TP}+\text{TN}}{n} = \frac{n_{11}+n_{22}}{n}. \end{aligned} \tag{3.16}$$

Accuracy is sometimes referred to as the proportion of "agreement" between predicted and actual values. However, a key disadvantage of this measure is that it can be severely influenced by the dominant category in imbalanced datasets, where this category will severely influence the overall accuracy score.

3.4.7.2 Precision or Positive Predictive Value

Precision estimates the probability of the true prediction given that a test or prediction is positive. In statistics and biostatistics it is called Positive Predictive Value (PPV). In this context, PPV/precision is used to describe the accuracy of diagnostic tests in when a test is positive. It is given by

$$\text{Precision} = \Pr(\text{Correct Predictions}|\text{Positive Prediction}) = \frac{\text{TP}}{\text{TP} + \text{FP}} = \frac{n_{12}}{n_{\bullet 2}},$$

where $n_{\bullet 2} = n_{12} + n_{22}$.

3.4.7.3 Sensitivity or recall

Sensitivity (also known as Recall in machine learning) is the proportion of correct prediction among the actual positive values and estimated the corresponding conditional probability of finding the truth among actual positive values.

$$\text{Sensitivity (Recall)} = \Pr(\text{Correct Prediction}|\text{Positive}) = \frac{\text{TP}}{\text{TP} + \text{FN}} = \frac{n_{22}}{n_{2\bullet}},$$

where $n_{2\bullet} = n_{21} + n_{22}$. Although in most cases we wish to calculate the Positive Predictive Value (PPV or Precision) or the corresponding Negative Predictive Value (NPV) since it a more useful measure this is not always feasible, particularly in biostatistics. The reason for this is that its calculation requires analyzing two groups: one for positively predicted individuals and one for negatively predicted individuals and In medical research, especially when dealing with relatively rare diseases, the positively predicted individuals (numerator in PPV/precision) is often very low due to the small prevalence of the disease. Therefore, a prospective study to collect data of this type would require a long time to gather a sufficient number of positively predicted data. On the other hand, sensitivity can be more obtained in a straightforward manner through a retrospective case-control study. Specifically, the data for the case (disease) group, which is harder to be found in the general population, are directly collected from hospitals (where these cases are reported), making the calculation of sensitivity feasible and easy. In predictive models, this option (increasing the case group) is not available, so using either sensitivity/recall or precision/PPV becomes equally valuable depending on the interpretation we seek in every problem.

3.4.7.4 Specificity

Specificity is given by the proportion of true negatives over all actual negative values and estimates the probability of predicting the truth given a negative response. It is paired with specificity and is used instead of the Negative Predictive Value (NPV) in order to be calculated from retrospective case-control studies. It is given by

$$\text{Specificity} = \Pr(\text{Correct Prediction}|\text{True Negative}) = \frac{\text{TN}}{\text{TN} + \text{FP}} = \frac{n_{11}}{n_{1\bullet}},$$

where $n_{1\bullet} = n_{11} + n_{12}$.

3.4.7.5 F1 score

F1 score combines the Precision (PPV) and recall (sensitivity) by considering their geometric mean. Hence, it provides a single metric that balances the trade-off between these two measures. It is particularly useful when there is an uneven class distribution or when both false positives and false negatives are important to consider. It is given by

$$\text{F1 Score} = \left[\frac{1}{2}\left(\frac{1}{\text{Precision}} + \frac{1}{\text{Recall}}\right)\right]^{-1} = 2 \times \frac{\text{Precision} \times \text{Recall}}{\text{Precision} + \text{Recall}}.$$

3.4.7.6 Summary for predictive measures

To conclude with, the accuracy offers a more basic, naive sometimes, approach to quantifying the efficiency of the model without considering any possible (and usually common) imbalance between the two predicted categories. Precision focuses more in the probability of identifying the truth among positive outcomes while sensitivity and specificity provide a more detailed picture of the model's effectiveness for both positive and negative responses.

3.4.7.7 Implementation in football

The implementation of these measures in football, where the number of final outcome categories are three instead of two, is not straightforward. In this occasion the confusion table will take the form of Table 3.4.

TABLE 3.4
Typical structure of a confusion matrix in football

Actual Outcome	Y	Predicted Outcome ($\hat{Y}$) Home wins ($\hat{Y}=1$)	Draw ($\hat{Y}=2$)	Home looses ($\hat{Y}=3$)
Home wins	$Y=1$	n_{11}	n_{12}	n_{13}
Draw	$Y=2$	n_{21}	n_{22}	n_{23}
Home looses	$Y=3$	n_{31}	n_{32}	n_{33}

There is one obvious approach to deal with the problem that in football we have three categories instead of two. One could ignore the draws and focus solely on the comparison between winning and losing games (from the perspective of the home team); however, this approach would fail to identify the model's poor prediction performance for draws. This can be complemented by a second set of predictive measures that focus on draws versus the other two outcomes. The problem with considering this additional set of predictive measures is that we do not have a single unified value for each measure making difficult to reach an overall conclusion or compare alternative predictive methods or models. Summary statistics, such as the mean of the two measures, can be considered, but we believe this is not good practice. The authors prefer to use Cohen's Kappa to estimate the agreement between different measurements or judges, which is appropriate for multicategory outcomes and is described in Section 3.4.8, which follows.

3.4.8 Cohen's Kappa for measuring agreement

Cohen's (1960) Kappa is a statistical measure used to evaluate the level of agreement between two raters (or judges) when classifying items into categories. It can be applied to both binary and multicategorical variables and is also useful for comparing actual and predicted outcomes, where the truth and predicted classifications play the role of ratings obtained by different raters or judges. It can be used as a substitute for the accuracy measure, which provides the proportion of agreement between actual and predicted outcomes but in a raw, unfiltered manner. The key difference with Kappa is that it adjusts for the agreement that could occur by chance, assuming independence between the two raters. Thus, unlike accuracy, which simply measures the proportion of agreement, Cohen's Kappa accounts for the possibility of random agreement, offering a more reliable metric for evaluating the predictive performance of a model.

Mathematically, κ will be calculated from a confusion matrix similar to 3.4. Hence, for an response variable with K categories, Cohen's Kappa (κ) is given as:

$$\kappa = \frac{P_o - P_e}{1 - P_e}, \text{ where } P_o = \frac{1}{n}\sum_{j=1}^{J} n_{jj} \text{ and } P_e = \frac{1}{n}\sum_{j=1}^{J} n_{j\bullet}n_{\bullet j}$$

where P_o = Accuracy is the observed agreement portion, and P_e is the expected agreement by chance (under the assumption of independence between actual and predicted outcomes), which is computed based on the marginal frequencies of the actual ($n_{j\bullet}$ for $j = 1, \ldots, J$) and predicted ($n_{\bullet j}$ for $j = 1, \ldots, J$) outcomes. Hence, we can rewrite κ as

$$\kappa = \frac{Accuracy - P_e}{1 - P_e} = 1 - \frac{1 - Accuracy}{1 - P_e}.$$

Concerning the interpretation of the values of kappa, it is clear that for $\kappa = 1$ then we have a method with perfect prediction, for $\kappa = 0$ then we have random prediction (similar to selection randomly the predicted outcome) and for $\kappa < 0$ then we have prediction which is even worse than random prediction. prediction In terms of scale, following the empirical characterization suggested by Landis and Koch (1977), for values lower than zero we have prediction worse than pure chance, for values between 0 and 0.2 we have low predictive performance, for values between 0.21 and 0.4 fair predictability, for values in 0.41–0.6 moderate predictability, for values in 9.61–0.8 substantial predictability and for values greater than 0.81 almost perfect predictive performance.

Cohen's Kappa provides a robust measure for raters agreement and inter-rater reliability that can be also used to measure the predictive ability of a method or a model by measuring the agreement between the predicted and the actual outcomes. It can be directly used for the 3×3 confusion matrix (see Table 3.4) obtained in football. It accounts for the random agreement or predictability. It is particularly valuable when evaluating the predictive performance of a model when the response outcome is a categorical variable with multiple categorical outcomes and in football.

3.5 Summary and closing remarks of Chapter 3

This chapter presented in detail the computational methods and model needed for tournament and game prediction, emphasizing simulation-based approaches and their practical applications in sports analytics.

We presented the use of predictive models for individual games, leveraging point estimates, bootstrap methods, and Bayesian approaches via MCMC. These techniques enable the generation of game scores, outcomes, and probabilities with varying degrees of precision and computational demand.

Monte Carlo simulations, both plug-in and bootstrap-based, are highlighted for their flexibility in regenerating leagues or tournaments. This approach facilitates an evaluation of model performance and provides a robust framework for exploring league scenarios and rankings. Similarly, Bayesian prediction methods, using posterior distributions, offer a probabilistic framework for quantifying uncertainties and enhancing predictive accuracy.

The chapter also examines league and tournament prediction scenarios, focusing on round-robin and hybrid formats. Predictive methodologies are tailored to specific challenges, such as mid-season forecasts and knock-out phase simulations. These scenarios illustrate the adaptability of statistical and computational tools in addressing diverse tournament structures and phases.

Performance assessment metrics, such as RMSE, MAE, and Brier scores, are introduced to evaluate the goodness-of-fit and predictive capabilities of models. The inclusion of ranked probability scores and pseudo-R^2 further strengthens the toolkit for assessing model accuracy and calibration. These measures underscore the importance of not just fitting the data but also ensuring out-of-sample predictive reliability.

Although the methods presented in this chapter provide a holistic framework for predictive modelling in sports tournaments, several opportunities for advancement remain. Incorporating dynamic covariates, such as player injuries, team strategies, or real-time form changes, could enhance model relevance. Additionally, integrating machine learning techniques with traditional statistical methods may yield superior predictive capabilities.

Future research could explore extending these methodologies to other sports or domains requiring sequential or hierarchical decision-making. Expanding simulation techniques to account for evolving tournament formats and incorporating real-time updates would further broaden the applicability of these models.

In summary, this chapter not only highlighted the current state of statistical modelling in tournament predictions but also set a clear pathway for future advancements. By combining robust methodologies with innovative applications, these models can continue to improve predictive accuracy and decision-making in sports analytics.

Next chapter moves to the implementation of basic models using the footbayes package providing a solid guidance of implementation to football data for the practitioner.

Appendix: Notation

Indexes and basic constants

- n: Number of games in the league/dataset
- K: Number of teams in the league/dataset
- T: Number of weeks in the league/dataset
- J: Number of levels/categories in a categorical variable; $J = 3$ for football match outcome.
- $i \in \{1, \ldots, n\}$: Observation/game index for game-arranged data
- $\ell \in \{1, 2\}$: Index denoting the home or away team for values one or two, respectively (for game-arranged data)
- $k \in \{1, \ldots, K\}$: Team index
- $w \in \{1, \ldots, W\}$: Week index
- $\imath$: observation index for univariate-arranged data with $\imath = 2i - 2 - \ell$
- $t \in \{1, \ldots, T\}$: Monte Carlo iteration index/superscript
- $j \in \{1, \ldots, J\}$: level/category index for a categorical variable

Model parameters

- $\boldsymbol{\theta}_{i1}, \boldsymbol{\theta}_{i2}$: parameter vectors for home and away teams
- ρ_i: dependence parameter between home and away goals
- μ: constant parameter in the vanilla model
- *home*: home effect parameter
- att_k, def_k: Fixed attacking and defensive parameter of k team
- $att_{k,t}$, $def_{k,t}$: Dynamic/random attacking and defensive parameter of k team at week t
- σ_a^2 and σ_d^2: random abilities variances
- $\beta_j^{(\ell)}$: effect of covariate j of home ($\ell = 1$) or away team ($\ell = 2$)

Variables and data for game-arranged data

- Y_{i1}, Y_{i2}: goals of the home and away team for game i
- $Z_i = Y_{i1} - Y_{i2}$: goal difference for game i
- Ω_i: Outcome of i game with three possible values: –1:home win, 2:draw, 3:away win; $\Omega_i = \mathcal{I}(Z_i > 0) + 2\mathcal{I}(Z_i = 0) + 3\mathcal{I}(Z_i < 0)$.
- $O_i = 2 - \Omega_i$: Outcome of i game with three possible values: -1: away win, 0: draw, 1:home win.
- $\lambda_{i\ell}$: Expected goals (in Poisson) of the home and away team for game i
- $\eta_{i\ell}$: Predictor of the home and away team for game i
- $X_{ij}^{(\ell)}$, $x_{ij}^{(\ell)}$: Covariates/Features for the home or away team

Variables and data for univariate-arranged data

- $Y_\imath$: goals scored in $\imath$ observation of +++ data; Note that $\imath = 2i - 2 - \ell$ hence $Y_\imath$ refers to the goals scored by the home team ($\ell = 1$) or the away team ($\ell = 2$) in game i

- $HT_\imath$, $AT_\imath$: covariates denoting the home and away teams
- $Home_\imath$: dummy variable denoting if the goals $Y_\imath$ were scored by a home team
- $X_{\imath j}^{(\ell)}$: Covariates/Features for the scoring team ($\ell = 1$) of $Y_\imath$ or the opponent team ($\ell = 2$) which receives the goals $Y_\imath$
- $Att_\imath$: attacking team which scores $Y_\imath$ goals
- $Def_\imath$: defending team which receives $Y_\imath$ goals

4

Implementation of basic models in R *via* footBayes

4.1 The installation of the footBayes package

Before starting with its functionalities, the footBayes R package (version 1.0.0) should be installed from the CRAN public repository (`https://CRAN.R-project.org/package=footBayes`) as explained in Code Snippet 6.

Code Snippet 6 footBayes package installation.

```
install.packages("footBayes")
library(footBayes)
```

Alternatively, the installation of the package is available also from the GitHub platform—nowadays, this source represents a standard practice for the release and the dissemination of R packages—through the following instructions in Code Snippet 7.

Code Snippet 7 footBayes package installation via GitHub.

```
library(devtools) # required
install_github("leoegidi/footBayes")
```

In order to familiarize with its extended use and appreciate the main functionalities, A thorough vignette accompanying the use of the package is available from the official CRAN package page at the link: `https://cran.r-project.org/web/packages/footBayes/vignettes/footBayes_a_rapid_guide.html`. To effectively use footBayes there is no need of any prior installation of other related packages; however, as explained in the next sections, the package strongly relies on the libraries/packages: rstan (Stan Development Team, 2022), which implements a robust Hamiltonian Monte Carlo (HMC) sampling (Betancourt, 2017) enginery within a Bayesian approach—for further details, see Chapter 2, Section 2.5.1.3—ggplot2, and dplyr.

DOI: 10.1201/9781003186496-4

4.2 Available models

The footBayes package is an encompassing modelling protocol that supports the fit of the following models for the number of goals/scores:

- double Poisson
- bivariate Poisson
- diagonal-inflated bivariate Poisson,

  and these other models for the goal difference:

  - Skellam
  - zero-inflated Skellam
  - student-t.

All these models allow a static and a dynamic fit—see Section 4.5.2 for further details—however, static models can be estimated according to either maximum likelihood (MLE) or Bayesian HMC methods, whereas dynamic estimation is available through HMC only.

4.3 Basic syntax and functions

footBayes is a very immediate and user-friendly R package designed to fit the most well-known football statistical models by typing just one line of code.

There are two main "fitting" functions:

- stan_foot(data, model, predict, ranking, dynamic_type, prior_par, home_effect): provides a "stanFoot" class object through Bayesian estimation based on the Stan ecosystem and Hamiltonian Monte Carlo (HMC) sampling. data should be a data-matrix or a data-frame containing the following mandatory items; periods, home_team, away_team, home_goals, away_goals. The user must be aware that a wrong inclusion, or a wrong permutation, of the quantities mentioned above cannot be restored by any of the other function options: thus, we suggest the user to double check the data before giving them as input for the fitting function. The model can be one among "double_pois", "biv_pois", "diag_infl_biv_pois", "skellam", "zero_infl_skellam" and "student_t". predict is an optional argument to specify the number of test-set/out-of-sample/held-out matches (if omitted, the test-set size is zero, and all the matches of the dataset are used to train the model). ranking is an optional "btdFoot" class argument or a data frame containing ranking points for teams with the following columns: periods, the time periods corresponding to the rankings (integer ≥ 1); team, the team names matching those in data (character string); and rank_points, the ranking points for each team (numeric). As an example for rankings, one could consider for instance the Coca-Cola rankings [1] typically adopted for the national teams in international competitions, such as the Euro or the World Cup. The user can specify to fit a dynamic model through the dynamic_type argument, whose possible choices are "weekly" and "seasonal". The user should be aware that the computational times arising from the two dynamic choices can be dramatically different: in fact, fitting weekly strengths' parameters, basically a batch for each of a league match day, is much more expensive than fitting a batch of seasonal parameters, one for each of the considered seasons. If dynamic_type is omitted, a static fit for the selected model is provided. Then, the user can specify different prior distributions' options through the argument prior_par, a list specifying the probability distributions for the parameters of interest, such as the team-specific abilities (ability), team-specific standard deviations (ability_sd), and home-effect (home). The possible choices for the team-specific abilities are "normal", "student_t", "cauchy", "laplace", analogously as the distribution names in the rstanarm package. Finally, the argument home_effect allows to specify a home-effect parameter, which is common in football modelling: by default this argument is set to "TRUE" for domestic national leagues, such as the Serie A, the Premier League, etc., however the user could set the argument to "FALSE", plausible when neither of the competing teams plays at its home stadium, as usually happens in international competitions such as Euro, World, or America's cups. Optional arguments related to the HMC sampling can be passed through a list, such as the number of HMC iterations (iter), or the number of Markov chains (chains). We refer the user to the stan function of the rstan library for further optional details.

- mle_foot(data, model, predict, ...): provides a "list" class object through maximum likelihood estimation, allowed for static models only. For the arguments data and predict we refer the user to the stan_foot function. For the argument model, the possible models are "double_pois", "biv_pois", "skellam", and "student_t". The user is then free to add some optional arguments for the maximum likelihood computation, such as the desired confidence intervals: interval = "profile" is the default option to calculate profile-likelihood confidence intervals, whereas interval = "Wald" can be specified to calculate Wald-type confidence intervals. Through the argument hessian = "TRUE" the user can obtain the computation of the Hessian matrix during the optimization procedure (default is "FALSE"). The argument method can be used to select one among the available optimization algorithms as provided by the base R function optim: by default, method = "BFGS", the quasi-Newton algorithm is chosen.

[1] `https://www.fifa.com/fifa-world-ranking`

NOTE: this function will be deprecated in future versions of the current package, when the incorporation of the CmdStan software (Gabry et al., 2024)—a wrapping Stan ecosystem—will allow to fit a single model written in Stan's syntax according to different algorithms, such as HMC, MLE, penalized likelihood, variational inference methods, Laplace's approximation.

Once the model has been fitted, the user may accomplish different tasks related to the following steps:

- *Teams' abilities visualization*: foot_abilities(object, data, type, teams) depicts the estimated attack and defence abilities along with a confidence/credible interval for both static and dynamic Poisson-based models. Red curves denote the attacking strengths, whereas blue curves denote the defensive strengths. The function yields some global ability measures for student-t models along with confidence/credible intervals—see Section 5.1.4 in Chapter 5. object is the fitted model object as estimated by either mle_foot or stan_foot; data is the original dataset; type is one among the following choices: c("attack", "defense", "both"), where by default "both" is selected. Finally, teams is a vector or a single string of valid team names for which the strength's estimates should be plotted.
- *model checking*: pp_foot(data, object, type, coverage) depicts two types of posterior-predictive checks (Gelman et al., 2013). type = "aggregated" returns a *cloud-plot* where for each observed frequency of the goal difference $\{-3, -2, -1, 0, 1, 1, 2\}$ the overlapped posterior-predictive distribution is depicted (default option). When type = "matches" is selected, the ordered goal differences for all the matches are displayed against the goal differences replicated from the posterior predictive distribution. The argument coverage controls the desired width of the credibility intervals when the argument type = "matches" is selected, and by default coverage = 0.95.
- *prediction for future matches*: foot_prob(data, object, home_team, away_team and foot_round_robin(data, object, teams) depict posterior predictive probabilities for a football season in a *chessboard-plot* and in a *round-robin* format, respectively, if the argument predict in the stan_foot function is explicitly specified (integer ≥ 0). The first function also provides a table for the home, draw and away winning probabilities. foot_rank(data, object, teams, visualize) yields rank-league reconstruction plots for in-sample matches or for out-of-sample matches. The user can choose among visualize = "aggregated" for a unique aggregated plot with credibility intervals for the final number of cumulated points, or visualize = "individual" to obtain the cumulated points for each team separately. We warn the user that these functions work only by passing a "stanFoot" object.
- *Ranking estimation via Bradley-Terry-Davidson models*: btd_foot(data, dynamic_rank, home_effect, prior_par, rank_measure) fits a Bayesian Bradley-Terry-Davidson model (Bradley and Terry, 1952; Davidson, 1970)—see

Chapter 1, Section 1.4 for further details— using the underlying Stan ecosystem, and supports both static and dynamic ranking models, allowing for the estimation of team strengths over time. As for the stan_foot function, the user can choose a dynamic estimation (dynamic_rank) and whether including an home effect parameter (home_effect); through the argument prior_par one may specify the prior distributions for: the team log-strengths (logStrength), the tie parameter (logTie), and the home-effect parameter (home), which is set to "FALSE" by default. rank_measure is a character string specifying the method used to summarize the posterior distributions of the team strengths, with one among "median", "mean", or "map". An object of the class "btdFoot" can also be given as an input for the ranking argument of stan_foot when a ranking covariate is desired. Thus, this function could be also used before launching the estimation procedure through stan_foot.

4.4 Basic models in footBayes

4.4.1 Double Poisson

As explained in Chapter 2, Sections 2.1.1–2.1.2, many scholars in the literature assume that the number of goals scored by each team follows a Poisson distribution (Maher, 1982; Lee, 1997; Rue and Salvesen, 2000; Baio and Blangiardo, 2010; Groll and Abedieh, 2013; Egidi et al., 2018b). Mathematically speaking, the joint distribution for the pair (Y_1, Y_2) is then given by the product of two Poisson probability functions:

$$
\begin{aligned}
fY_1, Y_2(y_1, y_2) = \Pr(Y_1 = y_1, Y_2 = y_2) = \\
\frac{\lambda_1^{y_1} \exp\{-\lambda_1\}}{y_1!} \frac{\lambda_2^{y_2} \exp\{-\lambda_2\}}{y_2!},
\end{aligned}
\tag{4.1}
$$

with means λ_1 and λ_2.

4.4.1.1 Model specification

To specify a proper statistical model, one could assume that the pair of random goals (Y_{i1}, Y_{i2}) is modelled as two *conditionally* independent Poisson random variables, for the i-th match, $i \in \{1, \ldots, n\}$. Analogously as in Equations (2.2) and (2.3), the double Poisson vanilla model takes then the general form:

$$
\begin{aligned}
Y_{i1}|\lambda_{i1} &\sim Poisson(\lambda_{i1}), \\
Y_{i2}|\lambda_{i2} &\sim Poisson(\lambda_{i2}), \\
\log(\lambda_{i1}) &= \mu + home + att_{h_i} + def_{a_i}, \\
\log(\lambda_{i2}) &= \mu + att_{a_i} + def_{h_i},
\end{aligned}
\tag{4.2}
$$

where $\lambda_{i1}, \lambda_{i2}$ represent the *scoring rates*, i.e. the expected number of goals for the home and away team, respectively; the parameters att_k and def_k encapsulate the offensive (or attacking) and defensive performances of team k, respectively, for each team k, $k \in \{1, \dots, K\}$; the nested indexes $h_i, a_i = 1, \dots, K$ denote the home and the away team playing in the i-th game, respectively; μ represents a constant parameter; *home* represents the *home-effect*, i.e. the well-known advantage of the team hosting the game. As a matter of interpretation, suppose that two teams of approximately equal strengths, with the further characteristic that the offensive ability of one team equals in absolute value the defensive ability of the opposing team, play one against the other, which means that $att_{h_i} + def_{a_i} \approx att_{a_i} + def_{h_i} \approx 0$; in such a case, the scoring rates in (4.2) are then influenced by the parameters μ and *home* only: if we provide some estimates for μ and *home* say equal to 0.2 and 0.34, the expected number of goals for the two competing teams will be approximately given by $\hat{\lambda}_{i1} = \exp\{0.2 + 0.34\} \approx 1.72$ and $\hat{\lambda}_{i2} = \exp\{0.2\} = 1.22$, respectively.

As suggested by Maher (1982) and Dixon and Coles (1997), a comfortable reparameterization for the log-linear scores in (4.2) is given by:

$$\begin{aligned} \lambda_{i1} &= \delta\gamma\alpha_{h_i}\beta_{a_i} \\ \lambda_{i2} &= \delta\alpha_{a_i}\beta_{h_i}, \end{aligned} \tag{4.3}$$

where $\delta = \exp\{\mu\}$, $\gamma = \exp\{home\}$, $\alpha_{h_i} = \exp\{att_{h_i}\}$, $\beta_{h_i} = \exp\{def_{h_i}\}$, $\alpha_{a_i} = \exp\{att_{a_i}\}$, and $\beta_{a_i} = \exp\{def_{a_i}\}$. The likelihood for the model (4.2) takes then the following form:

$$\begin{aligned} \mathcal{L}(\boldsymbol{\alpha}, \boldsymbol{\beta}, \gamma, \delta; y_1, y_2) &= \prod_{i=1}^{n} \lambda_{i1}^{y_{i1}} \exp\{-\lambda_{i1}\} \lambda_{i2}^{y_{i2}} \exp\{-\lambda_{i2}\} \\ &= \prod_{i=1}^{n} (\delta\gamma\alpha_{h_i}\beta_{a_i})^{y_{i1}} \exp\{-(\delta\gamma\alpha_{h_i}\beta_{a_i})\} \times \\ &\quad (\delta\alpha_{a_i}\beta_{h_i})^{y_{i2}} \exp\{-(\delta\alpha_{a_i}\beta_{h_i})\}. \end{aligned} \tag{4.4}$$

The log-likelihood is then:

$$\begin{aligned} \ell(\alpha, \beta, \gamma, \delta; y_1, y_2) = &\sum_{i=1}^{n} y_{i1} \log(\delta\gamma\alpha_{h_i}\beta_{a_i}) - \delta\gamma\alpha_{h_i}\beta_{a_i} + \\ &y_{i2} \log(\delta\alpha_{a_i}\beta_{h_i}) - \delta\alpha_{a_i}\beta_{h_i}. \end{aligned} \tag{4.5}$$

The model (4.2) has $2K + 2$ parameters. We note that many references in the literature do not explicitly include a constant intercept μ in the log-linear models for λ_{i1} and λ_{i2}. However, the inclusion of μ, despite not mandatory, is needed for identifiability purposes. In fact, assume that two approximately equal teams, as those considered above, play on a neutral pitch—this is the typical case occurring in World, Euro/Africa/America/Asia Cups, as remarked in Chapter 6 of this book—which means that the home-effect is set equal to zero. Then, we assume again that the offensive and defensive estimated skills compensate one each other, being $att_{h_i} + def_{a_i} \approx att_{a_i} + def_{h_i} \approx 0$. In this scenario, the overall intercept μ completely specifies λ_{i1} and λ_{i2}: by removing this parameter we would get in fact $\log(\lambda_{i1}) \approx \log(\lambda_{i2}) = 0$, then $\lambda_{i1} \approx \lambda_{i2} \approx 1$, which would make the resulting model non-identifiable. Moreover, in such scenarios where two teams are almost equal and with compensating skills, removing the common intercept μ has a direct implication: the draw 0-0 cannot occur, and this is not realistic in football, where 0-0 is the starting result and is not uncommon as the final result at all, especially in domestic leagues, such as the Italian Serie A, characterized by a low amount of scores.

Many variants and extensions of the general vanilla model form (4.2) have been proposed in the literature. Moreover, many computational methods are available to fit the model and provide reliable parameters' estimates, such as MLE and Bayesian methods.

4.4.1.2 The first Poisson-based model: Maher (1982)

Maher (1982), one of the first and more influential references for the modern football modelling focused on the Poisson distribution, is a particular case of model (4.2), with $\mu = home = 0$. The first version of the Maher's model allowed for distinct offensive and defensive parameters depending on whether the teams played at home or away, for a total of $4 \times K$ model parameters; however, some goodness-of-fit checks drove the conclusion that distinct attacking and defensive abilities when playing at home and away were unnecessary, and for this reason the number of offensive/defensive parameters in the general vanilla model (4.2) is usually set to $2 \times K$. One of the main novelties of the Maher's formulation regards the maximum likelihood parameters' estimation through Newton-Raphson iterative methods. According to the likelihood specification in Equations (4.5), one could get the MLEs for the home offensive/defensive parameters as follows:

$$\hat{\alpha}_k = \frac{\sum_{i:h_i=k} y_{i1}}{\sum_{i:a_i \neq k} \hat{\gamma}\hat{\delta}\hat{\beta}_i}, \quad \hat{\beta}_k = \frac{\sum_{i:h_i=k} y_{i2}}{\sum_{i:a_i \neq k} \hat{\delta}\hat{\alpha}_i}, \tag{4.6}$$

and an analogous procedure may be used for $\hat{\alpha}_{a_i}, \hat{\beta}_{a_i}$.

4.4.1.3 The Bayesian variant: Baio and Blangiardo (2010)

The double Poisson model proposed by Baio and Blangiardo (2010) is basically the same as in (4.2), except for the fact that $\mu = 0$. Moreover, the authors provide a Bayesian estimation procedure by eliciting some noninformative prior distributions for the model's parameters. The home-effect parameter *home* is modelled as a fixed effect and assigned a wildly noninformative prior distribution, whereas team-specific parameters are considered as exchangeable from a common distribution governed by group-level hyperparameters σ^2_{att} and

σ^2_{def}:

$$\begin{aligned} home &\sim N(0, \sigma^2_{home}), \\ att_k &\sim N(\mu_{\text{att}}, \sigma^2_{\text{att}}), \\ def_k &\sim N(\mu_{\text{def}}, \sigma^2_{\text{def}}), \end{aligned} \tag{4.7}$$

where $N(\mu, \sigma^2)$ denotes as usual a Gaussian distribution with mean μ and variance σ^2—or standard deviation σ. σ_{home} is set to 10^4 by the authors. The model's hyperparameters are then assigned some noninformative priors:

$$\begin{aligned} \mu_{\text{att}}, \mu_{\text{def}} &\sim N(0, 10^8), \\ \sigma_{\text{att}}, \sigma_{\text{def}} &\sim \text{invGamma}(0.1, 0.1), \end{aligned} \tag{4.8}$$

where invGamma(α, β) denotes an inverse Gamma distribution with shape parameter α and rate parameter β. Alternatively, in line with Gelman (2006), a more popular prior distribution for the team-specific standard deviations is represented by:

$$\sigma_{\text{att}}, \sigma_{\text{def}} \sim \text{Cauchy}^+(0, 5), \tag{4.9}$$

where Cauchy$^+$ denotes the half-Cauchy distribution with support $[0, +\infty)$.

NOTE: the default prior distributions for the group-level standard deviations supported by the footBayes package are given by those in Equation (4.9).

Posterior parameters estimates from the joint posterior distributions of the parameters are computed by use of Markov Chain Monte Carlo methods (MCMC) (Robert and Casella, 2013), specifically through the Gibbs sampling algorithm and the WinBUGS software (Lunn et al., 2000) (Geman and Geman, 1984b)—see Chapter 2, Section 2.5.1.2. A possible concern in the model above is the large amount of shrinkage caused by the Gaussian priors in (4.7): the Bayesian model above is likely to shrink the attack and the defence abilities towards their prior grand means $\mu_{\text{att}}, \mu_{\text{def}}$, being then not able to discern between good, intermediate, and poor teams. For such a reason, the authors propose to use in place of (4.7) a mixture consisting of three non-central student-t distributions in order to account for three possible latent categories of the teams (low, medium, high).

4.4.1.4 Model's parameter constraints and interpretation

To achieve global model's identifiability (Gelman and Hill, 2006) in Equation (4.2), we need to impose some constraints, the so-called "sum-to-zero" (STZ) identifiability constraints for the attacking and defensive parameters, as explained in Baio and Blangiardo (2010):

$$\sum_{k=1}^{K} att_k = \sum_{k=1}^{K} def_k = 0. \tag{4.10}$$

To accomplish with the condition in (4.10), as remarked in Chapter 2 Karlis and Ntzoufras (2003) one could assume the "corner constraint" that the abilities for the first team are equal to the negative sum of the $K-1$ residual abilities:

$$att_1 = -\sum_{k=2}^{K} att_k, \; def_1 = -\sum_{k=2}^{K} def_k. \tag{4.11}$$

Baio and Blangiardo (2010) suggest to use an alternative parametrization, the "centered constraint" by introducing some auxiliary parameters att_k^*, def_k^* such that:

$$\begin{aligned} att_k &= att_k^* - \overline{att} \\ def_k &= def_k^* - \overline{def}, \end{aligned} \tag{4.12}$$

where $\overline{att}$ and $\overline{def}$ represent the averages of the offensive and defensive parameters skills across the K teams. It is easy to check that the "new" att and def in Equation (4.12) satisfy the general condition expressed in Equation (4.10).

Furthermore, another "corner" STZ constraint to guarantee the general condition in (4.10) is suggested by the same authors by imposing a *baseline* team whose abilities are set to zero, with the residual teams abilities incremental with respect to the baseline team:

$$\begin{gathered} att_1 = 0, \; def_1 = 0 \\ \sum_{k=2}^{K} att_k = 0, \; \sum_{k=2}^{K} def_k = 0. \end{gathered} \tag{4.13}$$

Alternatively, Dixon and Coles (1997) propose a similar constraint based on the reparameterization (4.3)

$$K^{-1}\sum_{k=1}^{K} \alpha_k = 1, \quad K^{-1}\sum_{k=1}^{K} \beta_k = 1. \tag{4.14}$$

The choice between alternative constraints is mainly due to interpretation issues. Corner constraints can be faster to run than other STZ constraints from a computational perspective; however, the interpretation of the resulting coefficients may appear less intuitive, being made with respect to the baseline team associated with "fixed" attacking and defensive strengths.

4.4.1.5 Implementation in footBayes

In footBayes the user can fit a vanilla double Poisson model, either by adopting the maximum likelihood or the Bayesian estimation approach, via the following code in Code Snippet 8 by using the italy data contained in the package, just specifying the data and the model.

Code Snippet 8 Double Poisson model in footBayes.

```
## Some data
data(italy) # available in the package
italy_2009 <- subset(italy[, c(2,3,4,6,7)], Season =="2009")
colnames(italy_2009) <- c("periods", "home_team", "away_team",
                          "home_goals", "away_goals") # rename columns

## MLE fit
dp_mle <- mle_foot(data = italy_2009,
                   model= "double_pois")

## Bayesian fit
dp_stan <- stan_foot(data = italy_2009,
                     model="double_pois")
```

4.4.2 Bivariate Poisson

One of the main concerns with the double Poisson model (4.2) relies on the fact that the assumption of (conditional) independence between the goals scored during a match by two competing teams could be unrealistic (see Section 2.8.1 in Chapter 2 for some foundational details). In team sports of invasion such as football, water-polo, handball, hockey, and basketball, where the objective is to invade the opponent's territory and to score a goal or a point, it is reasonable to assume that the two outcome variables are correlated since the two teams interact during the game. To visualize this natural correlation, consider, for instance, the realistic football scenario of the home team leading 1-0, when only ten minutes are left to play: the away team may then become more determined and produce many efforts to score in order to equalize within the end of the match. Or, as another realistic situation, consider when the home team leads with a large margin, say, 3-0, or 4-0: it is plausible its players will be relaxing a bit, not forcing for another score, while the opponent team could take advantage of this relaxation to score at least one goal. To this aim, a positive goal correlation due to a change in the behaviour of the team, or both the teams, could be captured by a further parameter in the modelling specification. Although the assumption of positive correlation between the goals is widely accepted for national and domestic leagues, such as Serie A, La Liga, Premier League, and so on, it may be questionable for matches between national teams, such as the Euro/World Cup qualifiers. We need in fact to note that in such frameworks it is not unusual that a small national team such as San Marino plays against some of the strongest and traditional football national teams, such as Germany or England, according to a round-robin format. In these matches, the scores arising during a match could be observed under negative correlation: in simple words, the more goals are scored by England, and the smaller is the probability for San Marino to score.

To get a clue of this scenario, we invite the interested reader to focus his/her attention to the match San Marino-England, played on November 17th, 1993 in San Marino and valid for the World qualifiers for USA '94. The English team needed to win with a margin of seven goals or more and at the same time hope for the defeat of the Dutch national team against Poland. However...San Marino scored after eight seconds!—and the scorer, Davide Gualtieri, became a local hero[2]. In some sense, the goal scored by San Marino pushed the English team to score and score more goals, and this is in support of positive correlation; however, every goal scored by England—the match finished 7-1 for England—somehow diminished the possibility of other goals for San Marino, which was already "satisfied" for one of the fastest and most incredible goals in the international football history. In these scenarios, where a negative correlation could be realistic, assuming a copula-based model as in McHale and Scarf (2011b) in place of a full parametric model could be a suitable choice.

In this section we explain how capturing a positive correlation between the number of the goals scored by two opponents: the two outcome variables follow a bivariate Poisson distribution (Kocherlakota and Kocherlakota, 2017), such that the marginal distributions of the scores are still Poisson, but the random variables are now dependent. Before introducing a proper football model in order to generalize the double Poisson model (4.2), we give the general definition of the bivariate Poisson distribution.

Consider random variables $X_r, r = 1, 2, 3$, which follow independent Poisson distributions with parameters $\lambda_r > 0$. Then, the random variables $Y_1 = X_1 + X_3$ and $Y_2 = X_2 + X_3$ jointly follow jointly a bivariate Poisson distribution $\text{BP}(\lambda_1, \lambda_2, \lambda_3)$, with joint probability function given by

$$\begin{aligned} f_{Y_1,Y_2}(y_1, y_2) &= \Pr(Y_1 = y_1, Y_2 = y_2) \\ &= \exp\{-(\lambda_1 + \lambda_2 + \lambda_3)\}\frac{\lambda_1^{y_1}}{y_1!}\frac{\lambda_2^{y_2}}{y_2!}\sum_{k=0}^{\min(y_1,y_2)}\binom{y_1}{k}\binom{y_2}{k}k!\left(\frac{\lambda_3}{\lambda_1\lambda_2}\right)^k. \end{aligned} \tag{4.15}$$

Marginally each random variable follows a Poisson distribution with $\text{E}(Y_1) = \lambda_1 + \lambda_3$, $\text{E}(Y_2) = \lambda_2 + \lambda_3$, and $\text{cov}(Y_1, Y_2) = \lambda_3$; λ_3 acts then as a measure of dependence between the goals scored by the two competing teams. If $\lambda_3 = 0$ then the two variables are conditionally independent and the bivariate Poisson distribution reduces to the product of two independent Poisson distributions, the double Poisson model (4.2). For a comprehensive treatment of the bivariate Poisson distribution and its multivariate extensions see Kocherlakota and Kocherlakota (2017) and Johnson et al. (1997).

Due to its flexibility, the bivariate Poisson distribution is then particularly appealing for modelling dependence in team sports: in terms of global interpretation, λ_1 and λ_2 represent, as for the double Poisson model, the "net" scoring strengths for each team, whereas λ_3 reflects game conditions, such as

[2]`https://www.youtube.com/watch?v=y4L38WmFuZo`

the speed of the game, the weather, the climate, the stadium, and so on. As suggested by Koopman and Lit (2015, 2019a), a higher λ_3 leads typically to a higher number of equal observations ($Y_1 = Y_2$) which for a football match is a draw. Maher (1982) argued that a match does not consist of two independent processes for the number of the scores, and found some estimates for λ_3 according to some frequentist goodness-of-fit tests. Also Dixon and Coles (1997) proposed an extension of the basic double Poisson model to capture scores' dependence and draw inflation—we refer to the next section for a quick illustration of their procedure. Karlis and Ntzoufras (2003) proposed a thorough specification of a bivariate Poisson model for the number of goals scored by the home and the away team and make use of the EM algorithm to obtain maximum likelihood parameter estimates.

4.4.2.1 Model specification

Then, the general form of a bivariate Poisson model takes the following form:

$$\begin{aligned}(Y_{i1}, Y_{i2} | \lambda_{i1}, \lambda_{i2}, \lambda_{i3}) &\sim \text{BP}(\lambda_{i1}, \lambda_{i2}, \lambda_{i3}) \\ \log(\lambda_{i1}) &= \mu + home + att_{h_i} + def_{a_i}, \\ \log(\lambda_{i2}) &= \mu + att_{a_i} + def_{h_i},\end{aligned} \tag{4.16}$$

where the specification for the log-linear scores is the same as in (4.2). For the covariance parameters λ_{i3} one may assume various versions of the linear predictor, we propose here the general form:

$$\log(\lambda_{i3}) = \psi_0 + \omega_1 \psi_{h_i} + \omega_2 \psi_{a_i} + \omega_3 \boldsymbol{\eta} U_i, \tag{4.17}$$

where ψ_0 is a constant parameter, ψ_{h_i} and ψ_{a_i} are parameters that depend on the home and away team respectively, U_i is a vector of covariates for the i-th match used to model the covariance term, and $\boldsymbol{\eta}$ is the corresponding vector of regression coefficients. The parameters ω_1, ω_2, and ω_3 are dummy binary indicators taking values 0 or 1 that are able to "activate" distinct sources of the linear predictor. Hence, when $\omega_1 = \omega_2 = \omega_3 = 0$ we consider constant covariance, whereas when $(\omega_1, \omega_2, \omega_3) = (1, 1, 0)$ we assume that the covariance depends on the teams' parameters only, but not on further match covariates. According to the formulation in (4.17), the parameter λ_{i3} can be interpreted as a random effect which acts additively on the marginal means and reflects game conditions.

The interpretation of the bivariate Poisson model in Equation (4.16) is similar to the double Poisson model in (4.2), except for the fact that an explicit positive correlation between the teams' scores is now introduced by (4.17). The influence of this parameter on the other estimated parameters will be broadly investigated in Section 4.5. To achieve model identifiability for (4.16), the same STZ constraints proposed for the double Poisson model in Section 4.4.1.4 could be applied for the offensive and defensive parameters here as well.

Suppose now to apply the same parametrization (4.3) adopted for the double Poisson model. For simplicity, we assume a constant covariance ($\omega_1 =$

$\omega_2 = \omega_3 = 0$) depending then only on ψ_0, and we further assume $\lambda_3 \equiv \epsilon = \exp\{\psi_0\}$. Then the likelihood of the model is given by

$$\begin{aligned}\mathcal{L}(\boldsymbol{\alpha}, \boldsymbol{\beta}, \gamma, \delta, \epsilon; y_1, y_2) &= \prod_{i=1}^{n} \lambda_{i1}^{y_{i1}} \exp\{-\lambda_{i1}\} \lambda_{i2}^{y_{i2}} \exp\{-\lambda_{i2}\} \exp\{-\lambda_{i3}\} \times \\ &\quad \sum_{k=0}^{\min(y_{i1}, y_{i2})} \binom{y_{i1}}{k} \binom{y_{i2}}{k} k! \left(\frac{\lambda_{i3}}{\lambda_{i1}\lambda_{i2}} \right)^k \\ &= \prod_{i=1}^{n} (\delta\gamma\alpha_{h_i}\beta_{a_i})^{y_{i1}} \exp\{-(\delta\gamma\alpha_{h_i}\beta_{a_i})\} \times \\ &\quad (\delta\alpha_{a_i}\beta_{h_i})^{y_{i2}} \exp\{-(\delta\alpha_{a_i}\beta_{h_i})\} \exp\{-\epsilon\} \times \\ &\quad \sum_{k=0}^{\min(y_{i1}, y_{i2})} \binom{y_{i1}}{k} \binom{y_{i2}}{k} k! \left(\frac{\epsilon}{\delta^2\gamma\alpha_{h_i}\beta_{a_i}\alpha_{a_i}\beta_{h_i}} \right)^k .\end{aligned} \tag{4.18}$$

The model (4.16)—(4.17) with $\omega_1 = \omega_2 = \omega_3 = 0$ has $2K+3$ parameters—one more than the model (4.2)—whereas the bivariate Poisson model with $\omega_1 = \omega_2 = 1, \omega_3 = 0$ has $4K+3$ parameters.

4.4.2.2 Implementation in footBayes

In footBayes the user can fit a bivariate Poisson model, either by adopting the maximum likelihood or the Bayesian estimation approach, via the following code contained in Code Snippet 9 just specifying the data and the model.

Code Snippet 9 Bivariate Poisson model in footBayes.

```
## MLE fit
biv_mle <- mle_foot(data = italy_2009,
                    model= "biv_pois")


## Bayesian fit
biv_stan <- stan_foot(data = italy_2009,
                      model= "biv_pois")
```

4.4.3 Dynamic models

A structural limitation from the previous models regards the assumption of static team-specific parameters, which means that the teams are assumed to have a constant performance across time, as determined by the offensive and defensive abilities. However, in football analytics it is realistic to assume that teams' performances, both in domestic leagues and international competitions, tend to be dynamic and change across different seasons, if not different weeks within the same season, or match-days within the same competition. As widely known, many distinct factors could contribute to the temporal variation in the performances, we list here just a bunch of them:

- the teams in the domestic leagues are usually involved in the summer/winter players' transfermarket, and this could dramatically change the quality and the composition of their rosters;
- some teams' players could be injured in some periods and this could affect the quality of the teams they belong to;
- the football coaches could be dismissed from their teams due to some non-effective results or other related problems;
- some teams could improve/worsen their attitudes due to the so-called turnover;
- some teams could under/over perform in some periods due to their fitness conditions;

and many more. For these and other reasons, we explore here some dynamic extensions provided in the literature in terms of:

- altering the likelihood through a weighting function in order to acknowledge how the past historical information could contribute to the most recent performances (*indirect* method);
- specifying a time-dependent stochastic model for the offensive and defensive parameters in the Poisson-based models, or for the global abilities in the student-t model (*direct* method).

Many of the proposals in the literature combine both the direct and the indirect procedures to account for dynamic trends: we deal with a quick overview in the next sections.

4.4.4 Weighting function in the likelihood

Regarding the indirect method aforementioned, Dixon and Coles (1997) propose to modify Equation (4.4) by introducing a pseudo-likelihood for each time point t:

$$\mathcal{L}(\boldsymbol{\alpha}, \boldsymbol{\beta}, \gamma, \rho; y_1, y_2) = \prod_{i \in A_t} \{\tau_{\lambda_{i1}, \lambda_{i2}}(y_{i1}, y_{i2}) \lambda_{i1}^{y_{i1}} \exp\{-\lambda_{i1}\} \lambda_{i2}^{y_{i2}} \exp\{-\lambda_{i2}\}\}^{\phi(t-t_i)}, \quad (4.19)$$

where t_i is the time that match i was played, $A_t = \{i : t_i < t\}$, $\lambda_{i1}, \lambda_{i2}$ are defined as in Equation (4.3), τ is a scaling parameter accounting for correlation of low-scores results, and ϕ is a non-increasing function of time. For a thorough illustration of this *scaled double Poisson model* proposed by Dixon and Coles (1997) we refer to Chapter 5, Section 5.4.1. Actually, in Equation (4.19) the parameters α_k, β_k, ρ, and γ are themselves time-dependent; maximizing this equation at time t leads to parameters estimates which are based on games up to time t only. The choice of the function ϕ is crucial to give more or less emphasis to past historical information. Dixon and Coles (1997) propose to use:

$$\phi(t) = \exp(-\xi t), \tag{4.20}$$

for which all the previous results are exponentially downweighted according to a parameter $\xi > 0$. The static model likelihood (4.4) arises as a special case when $\xi = 0$, whereas larger values of ξ give more weight to the most recent results. We note that the choice/optimization of the weighting parameter ξ can be problematic: the authors propose to choose ξ in light of the values yielding the best predictive accuracy.

4.4.4.1 Truncated Poisson

Another indirect way to account for dynamic trends is introduced by Rue and Salvesen (2000) and Owen (2011), who propose a Bayesian dynamic double Poisson model to estimate the time-dependent skills of all the teams competing in a league, by using the powerful MCMC enginery to fit the model model and retrieve retrospective prediction. Rue and Salvesen (2000) modify the model likelihood (4.19) of Dixon and Coles (1997) in two directions, by using a truncated Poisson distribution in place of a canonical Poisson, and providing a greater amount of model's robustness. They propose to use an upper-truncated Poisson distribution with truncation set at five goals, since the goals each team scores after five are not much informative for the team-specific abilities. Moreover, they deem that only a fraction of the match results is worth to affect and influence the offensive and defensive abilities between two competing teams. They specify the following model for the number of scores:

$$\begin{aligned} f_{Y_1,Y_2}(y_1, y_2) = (1-\epsilon)\kappa_{\lambda_{i1},\lambda_{i2}}(y_{i1}, y_{i2})\mathcal{P}oisson^*(\lambda_{i1})\mathcal{P}oisson^*(\lambda_{i2}) \\ + \epsilon\kappa_{\lambda_{i1},\lambda_{i2}}(y_{i1}, y_{i2})\mathcal{P}oisson^*(e^{\mu_h})\mathcal{P}oisson^*(e^{\mu_a}), \end{aligned} \tag{4.21}$$

where $\mathcal{P}oisson^*$ denotes an upper-truncated Poisson distribution with upper truncation at five goals, λ_{i1} and λ_{i2} are the scoring rates as defined as in (4.1), the correlation factor κ behaves similarly as in Equation (4.19)—see Rue and Salvesen (2000) for further details—ϵ is the mixture parameter, such that only $100(1-\epsilon)\%$ of the information in the match result is informative concerning the teams' abilities; the non-informative part in (4.21) uses the average goal intensities $\exp(\mu_h), \exp(\mu_a)$—μ_h and μ_a denote global constants describing (roughly) the logarithm of the empirical mean of the home and away goals, respectively: compared to the general formulation in (4.2), one could note that the global intercept $\mu = (\mu_h, \mu_a)$ depends on whether the home or the away team is considered, see (4.22) below for other details. The bigger is ϵ and the more shrunken are the scores towards the low for an average match—the authors claim that the estimated value for ϵ is about 0.2.

4.4.4.2 Dynamic abilities

Moreover, in order to directly consider time evolution across the season, Rue and Salvesen (2000) and Owen (2011) specify time-dependent team-specific abilities for team $k, k = 1, \ldots, K$, in correspondence of the distinct match-times t, $t = 1, 2, \ldots, T$, denoted with $att_{k,t}, def_{k,t}$ for a generic $t \geq 0$. By extending (4.2), the scoring intensities can take then the general form:

$$\begin{aligned} \log(\lambda_{1i}) &= \mu_h + home + att_{h_i,t} + def_{a_i,t} \\ \log(\lambda_{2i}) &= \mu_a + att_{a_i,t} + def_{h_i,t}. \end{aligned} \tag{4.22}$$

4.4.4.3 Dynamic abilities with psychological effects

In order to further directly model the dynamic abilities, Equation (4.22) could be extended by including in this specification a so-called psychological effect:

$$\begin{aligned} \log(\lambda_{1i}) &= \mu_h + home + att_{h_i,t} + def_{a_i,t} - \zeta\Delta_{12,t} \\ \log(\lambda_{2i}) &= \mu_a + att_{a_i,t} + def_{h_i,t} + \zeta\Delta_{12,t}, \end{aligned} \tag{4.23}$$

where $\Delta_{12,t} = (att_{h_i,t} - def_{h_i,t} - att_{a_i,t} + def_{a_i,t})/2$ measures the difference in strength between the two teams at time t, and ζ is a further parameter giving the magnitude of the psychological effect. In this framework, we assume that the strengths of the two teams are not much different, since we usually focus on teams in the same league, so it is reasonable to expect $\zeta > 0$—as motivated by Rue and Salvesen (2000), the opposite effect, $\zeta < 0$ might occur if the home team is so superior compared with the away team that the latter develops an inferiority complex facing the home team, which we do not expect will happen in the same league.

4.4.4.4 Prior distributions and STZ constraints

To allow the team-specific parameters to vary over time in a Bayesian setting, Rue and Salvesen (2000) propose to use some particular prior distributions to tie together *att* and *def* at successive time points $t - 1$ and t. We highlight then the *auto-regressive prior distributions* for the offensive and defensive parameters:

$$\begin{aligned} att_{k,t} &\sim N\left(att_{k,t-1}, \sigma^2_{\text{att}}\right), \\ def_{k,t} &\sim N(def_{k,t-1}, \sigma^2_{\text{def}}), \end{aligned} \tag{4.24}$$

where the level of the abilities at time t is centred around the same skill value at time $t-1$, and $\sigma_{\text{att}}, \sigma_{\text{def}}$ reflect the evolution standard deviations for all the teams and, for the attacking and defensive abilities, respectively. Some STZ identifiability constraints as in (4.10) are assumed for each match-time t:

$$\sum_{k=1}^{K} att_{k,t} = 0, \ \sum_{k=1}^{K} def_{k,t} = 0. \tag{4.25}$$

We further assume prior distributions at time zero for the team-specific abilities representing baseline attack and defence strengths at the beginning of the season:

$$\begin{aligned} att_{k,0} &\sim N(\mu_{\text{att}}, \sigma^2_{0,\text{att}}), \\ def_{k,0} &\sim N(\mu_{\text{def}}, \sigma^2_{0,\text{def}}), \end{aligned} \tag{4.26}$$

where μ_{att} and μ_{def} are hyper-parameters representing prior means for these baseline attack and defence strengths, and $\sigma^2_{0,\text{att}}, \sigma^2_{0,\text{def}}$ are the prior variances at time zero (to inform these prior means, one could also follow an *informative Bayes* approach and fix the hyper-parameters with respect to some past team performance). To complete the model, some prior distributions for all the other model parameters are required, and inference from the posterior distribution could be obtained by use of MCMC methods, as suggested by the authors.

4.4.4.5 Gaussian processes for the dynamic abilities

Another alternative to directly capture the temporal trend in the offensive and defensive abilities is represented by the adoption of two Gaussian processes (Egidi et al., 2018a): given the times $t = 1, 2, \ldots, T$ with $t \in \mathbb{R}$, the prior probability of a finite number of attack/defence parameters conditioned on their inputs is then a multivariate Gaussian:

$$\begin{aligned} att_{k,\cdot} &\sim N_T(\mu_{k,att}(\cdot), K_{att}(\cdot)), \\ def_{k,\cdot} &\sim N_T(\mu_{k,def}(\cdot), K_{def}(\cdot)), \end{aligned} \tag{4.27}$$

where $\mu_{k,att}(\cdot), \mu_{k,def}(\cdot)$ are T-vectors, and $K_{att}(\cdot), K_{def}(\cdot)$ are two $T \times T$ covariance matrices. The mean functions $\mu : \mathbb{R}^T \to \mathbb{R}^T$ can be anything, but the covariance function $K : \mathbb{R}^T \to \mathbb{R}^{T \times T}$ must produce a positive-definite matrix for any input t. The authors provide a HMC procedure to estimate these parameters—see Section 2.5.1.3 in Chapter 2 for further details.

4.4.4.6 Autoregressive process for the team-specific abilities

Koopman and Lit (2015) extend the bivariate Poisson specification of Karlis and Ntzoufras (2003) in (4.16) towards a direct dynamic approach under maximum likelihood estimation. By maintaining the previous notation, the model specification is as follows:

$$\begin{aligned} (Y_{i1}, Y_{i2} | \lambda_{i1}, \lambda_{i2}, \lambda_{i3}) &\sim \text{BP}(\lambda_{i1}, \lambda_{i2}, \lambda_{i3}) \\ \log(\lambda_{i1}) &= \mu + home + att_{h_i,t} + def_{a_i,t} \\ \log(\lambda_{i2}) &= \mu + att_{a_i,t} + def_{h_i,t}. \end{aligned} \tag{4.28}$$

To take into consideration the time evolution in the teams' abilities, they propose an auto-regressive process for both the attack and the defence parameters:

$$\begin{aligned} att_{k,t} &= \gamma_k^{att} + \phi_k^{att} att_{k,t-1} + \eta_{k,t}^{att} \\ def_{k,t} &= \gamma_k^{def} + \phi_k^{def} def_{k,t-1} + \eta_{k,t}^{def}, \end{aligned} \tag{4.29}$$

where $\gamma_k^{att}, \gamma_k^{def}$ are unknown constants, $\phi_k^{att}, \phi_k^{def}$ are autoregressive coefficients, and $\eta_{k,t}^{att}, \eta_{k,t}^{def}$ are normally distributed error terms which are independent of each other for each team k and time t. The authors assume a stationary process for the team-specific abilities by requiring the conditions $|\phi_k^{\cdot}| < 1$. The independent disturbance sequences are stochastically generated by:

$$\eta_{k,t}^{\cdot} \sim N(0, \sigma_k^2), \tag{4.30}$$

whereas the initial conditions for the auto-regressive processes are based on means and variances of their unconditional distributions—here for the attack only:

$$\mathrm{E}(att_{k,t}) = \gamma_k^{att}/(1 - \phi_k^{att}), \quad \mathrm{Var}(att_{k,t}) = \sigma_k^2/(1 - (\phi_k^{att})^2). \tag{4.31}$$

4.4.4.7 Implementation in footBayes

In footBayes the user can fit Bayesian dynamic models by allowing team-specific abilities as in Equations (4.22) and (4.24) via the following code in Code Snippet 10, by using the dynamic_type argument in the stan_foot function. Note that the "weekly" option is appropriate when data relate to matches of the same season.

Code Snippet 10 Dynamic models in footBayes.

```
## dynamic Bayesian fit
dp_stan_dyn <- stan_foot(data = italy_2009,
                         model= "double_pois",
                         dynamic_type ="weekly")
biv_stan_dyn <- stan_foot(data = italy_2009,
                          model= "biv_pois",
                          dynamic_type ="weekly")
```

4.5 Case-study: Italian Serie A 2009/2010

In order to fully illustrate the main functionalities of the package, we import some data about the Italian Serie A, season 2009/2010, from the italy dataset available from the footBayes package. The season consists of $K = 20$ teams and $T = 38$ match-days teams according to a round-robin format, where the teams compete in a home-and-away format, playing against each other twice in total: once at their home stadium and once at their opponent's. We start with some basic Poisson-based models covered in Sections 4.4 by adopting:

- *the likelihood approach*: the mle_foot function returns a "list" class object containing the MLE estimates along with 95% profile-likelihood deviance confidence intervals (by default) or Wald-type confidence intervals. The user can choose the desired confidence interval with the optional argument interval = c("profile", "Wald").
- *the Bayesian approach*: the stan_foot function returns a "stanFoot" class object containing the results of an HMC posterior sampling performed through the underlying rstan ecosystem. Beyond the usual arguments, the user can eventually choose the number of iterations (iter), the number of Markov chains (chains), and other optional arguments values related to the sampling algorithm. Through the print function, the usual Bayesian model summaries can be obtained: posterior means, medians, standard deviations, percentiles at 2.5%, 25%, 75%, 97.5% level; moreover, the effective sample size (n_eff) and the Gelman-Rubin statistic (Rhat) (Gelman et al., 1992) to monitor the algorithm's convergence are available.

At this initial stage, we currently ignore any time-dependence in our parameters, considering them to be static across distinct match-days of the 2009/2010 season. We start by exploring the data, a data table consisting of five columns: the season (periods), the home team (home_team), the away team (away_team) the home goals (home_goals) and away goals (away_goals). Data acquisition and columns' renaming are reported in Code Snippet 11, whereas a summary of the data is shown in Output 7. Each row corresponds to the information for a single match.

Code Snippet 11 Italian Serie A 2009/2010: data acquisition.

```
### some required packages (install them!)

library(footBayes)
library(devtools)
library(dplyr)
library(bayesplot)
library(ggplot2)
library(loo)

### Use Italian Serie A 2009/2010

data(italy)  # available in footBayes
italy_2009 <- subset(italy[, c(2,3,4,6,7)], Season =="2009")
colnames(italy_2009) <- c("periods", "home_team", "away_team",
                          "home_goals", "away_goals") # rename columns
head(italy_2009)
```

```
      periods      home_team       away_team home_goals away_goals
17501    2009     Bologna FC ACF Fiorentina          1          1
17502    2009       AC Siena        AC Milan          1          2
17503    2009          Inter         AS Bari          1          1
17504    2009 Calcio Catania       Sampdoria          1          2
17505    2009      Genoa CFC         AS Roma          3          2
17506    2009       Juventus   Chievo Verona          1          0
```

Output 7: Italian Serie A 2009/2010 dataset: a summary. Each row denotes a single match.

4.5.1 Static models

We start by fitting the Bayesian double and bivariate Poisson models by using four parallel Markov chains and 2000 iterations for each chain. The home-effect is included since we are framed in a domestic leagues with home/away matches. Then, as shown in Code Snippets 12 and 13 and reported in Outputs 8 and 9, we print the posterior estimates for some selected parameters (here iterations messages and other eventual warnings/messages from the R consolle are suppressed to ease the readability).

Code Snippet 12 Italian Serie A 2009/2010: double Poisson model.

```
### Fit static Stan models

dp_stan <- stan_foot(data = italy_2009,
                     model="double_pois",
                     home_effect = TRUE)     # double poisson
print(dp_stan, pars =c("home", "sigma_att", "sigma_def"))
```

```
Summary of Stan football model
------------------------------

Posterior summaries for model parameters:
          mean se_mean   sd 2.5%  25%  50%  75% 97.5% n_eff Rhat
home      0.41       0 0.04 0.32 0.38 0.41 0.44  0.49  4779 1.00
sigma_att 0.20       0 0.06 0.10 0.16 0.19 0.23  0.32  1327 1.00
sigma_def 0.11       0 0.05 0.03 0.08 0.11 0.15  0.23   329 1.03
```

Output 8: Double Poisson model's summary from stan_foot. Posterior estimates for the selected parameters.

Code Snippet 13 Italian Serie A 2009/2010: bivariate Poisson model.

```
biv_stan <- stan_foot(data = italy_2009,
                      model="biv_pois",
                      home_effect = TRUE) # biv poisson
print(biv_stan, pars =c("home", "rho",  "sigma_att", "sigma_def"))
```

```
Summary of Stan football model
------------------------------

Posterior summaries for model parameters:
           mean se_mean   sd  2.5%   25%   50%   75% 97.5% n_eff Rhat
home       0.34    0.00 0.05  0.23  0.30  0.34  0.37  0.44  3782    1
rho       -2.36    0.01 0.42 -3.34 -2.60 -2.31 -2.06 -1.67  3395    1
sigma_att  0.20    0.00 0.06  0.10  0.16  0.20  0.24  0.33  1603    1
sigma_def  0.12    0.00 0.05  0.03  0.08  0.11  0.15  0.22   397    1
```

Output 9: Bivariate Poisson model's summary from Stan.

The Gelman-Rubin statistic $\hat{R}$ (Rhat) reported in the last column is below the threshold 1.1 for all the parameters in both the models in Outputs 8 and 9, respectively; the effective sample size (n_eff), measuring the approximate number of iid replications from the Markov chains, does not appear to be problematic. Thus, HMC sampling reached the convergence to the stationary distribution for all the parameters of interest.

As we could expect, the Bayesian fit suggests there is an estimated positive home-effect, being the posterior mean for home equal to 0.41 in the double Poisson and 0.34 in the bivariate Poisson model, which means that there is an estimated multiplicative effect of $\exp\{0.41\} \approx 1.51$ and $\exp\{0.34\} \approx 1.40$ in the average goals scored by the home team, respectively. As a matter of interpretation, if we consider the bivariate Poisson model this implies that if the abilities of the two teams do compensate one each other in such a way that the sum of one team's attack and the other team's defense is approximately equal to zero, then the average number of goals for the home team will be $\hat{\lambda}_1 = \exp\{0.34\} \approx 1.40$, against $\hat{\lambda}_2 = \exp\{0\} = 1$. The posterior estimates for the attack and defence standard deviations are basically the same in the two models.

NOTE: consider that in this model specification from stan_foot, unlike for what is explicitly specified in Equations (2.3) and (4.2), the global intercept μ

is not included. However, as a matter of modelling interpretation and global identifiability, here $\mu = \mu_{\text{att}} + \mu_{\text{def}}$, where μ_{att} and μ_{def} are the hyper-prior means of the team-specific abilities specified in Equation (4.7) for the Bayesian models. Thus, μ is not directly reported as a fixed-effect estimated term in Outputs 8 and 9, rather it is absorbed in the terms μ_{att} and μ_{def}.

In the bivariate Poisson model estimated above, we assume that the covariance λ_3 defined in Equation (4.17) is constant, thus it does not depend on the match or teams' characteristics, or further covariates:

$$\begin{aligned} \lambda_3 &= \exp\{\rho\} \\ \rho &\sim N^+(0,1), \end{aligned}$$

where ρ is assigned an half-Gaussian prior distribution with standard deviation equal to 1—in Equation (4.17) we used the symbol ψ_0 to denote the intercept in the covariance specification, here we use ρ to simplify the notation. According to the fit above, we get an estimate of $\lambda_3 = \exp\{-2.36\} \approx 0.094$, suggesting a low, despite non-null, amount of goal-correlation existing from the 2009/2010 Italian Serie A. In further package versions beyond the 1.0.0, the user will be allowed to specify a more general linear predictor for $\log(\lambda_{i3})$, as outlined in Equation (4.17) in Section 4.4.2, along with some prior distributions for the parameters involved in the covariance formulation.

In terms of visualization, we can depict the marginal posterior distributions for ρ, and eventually for the other *fixed-effects* parameters of the bivariate Poisson model, using the bayesplot R package designed for Bayesian visualizations, as shown in Code Snippet 14 and reported in Output 10.

Code Snippet 14 Italian Serie A 2009/2010: marginal posterior distributions.

```
## Marginal posterior with bayesplot

posterior1 <- as.matrix(biv_stan$fit)
mcmc_areas(posterior1, regex_pars=c("home", "rho",
                                    "sigma_att", "sigma_def"))+
  xaxis_text(on =TRUE, size=ggplot2::rel(2.9))+
  yaxis_text(on =TRUE, size=ggplot2::rel(2.9))+
  scale_y_discrete(labels = c("home", expression(rho),
                   expression(sigma[att]), expression(sigma[def])))+
  theme(plot.title = element_text(hjust = 0.5, size =rel(2.6)))
```

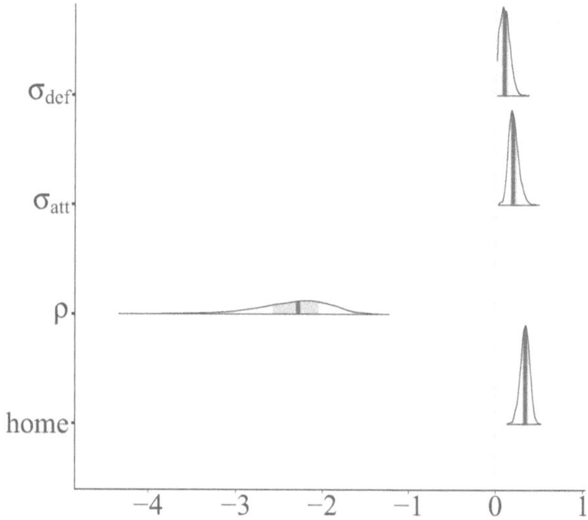

Output 10: Italian Serie A 2009/2010: posterior marginal distributions from the bivariate Poisson model (iter = 2000, four parallel chains). The dark grey lines denote the posterior medians, whereas the light grey area corresponds to 50% high posterior density (HPD) credible intervals.

We estimate now the double and bivariate Poisson models under the MLE approach with Wald-type confidence intervals and print the MLE estimates, e.g. for the parameters fixed-effect parameters, as shown in Code Snippets 15 and 16 and reported in Outputs 11 and 12.

Code Snippet 15 Italian Serie A 2009/2010: double Poisson model (ML) estimate.

```
### Fit static MLE models (wald-type CIs)

dp_mle <- mle_foot(data = italy_2009, model="double_pois",
                   interval = "Wald") # mle biv poisson
tab_dp <- rbind(dp_mle$home)
rownames(tab_dp) <- c("home")
tab_dp
```

```
     2.5%  mle 97.5%
home 0.31 0.39  0.47
```

Output 11: Maximum likelihood estimates for the double Poisson model using Wald-type confidence intervals.

Code Snippet 16 Italian Serie A 2009/2010: bivariate Poisson model (ML).

```
biv_mle <- mle_foot(data = italy_2009, model="biv_pois",
                    interval = "Wald") # mle biv poisson
tab_biv <- rbind(biv_mle$home, round(log(biv_mle$corr),2))
rownames(tab_biv) <- c("home", "rho")
tab_biv
```

```
       2.5%   mle 97.5%
home  0.25  0.34  0.42
rho  -2.53 -2.53 -2.41
```

Output 12: Maximum likelihood estimates for the bivariate Poisson model using Wald-type confidence intervals.

As we may notice from Outputs 11 and 12, MLE and Bayesian models give very similar results in terms of parameters' estimates, at least regarding the home effect and the correlation parameter, being the estimated home equal to 0.39 in the double Poisson and 0.34 in the bivariate Poisson model, whereas $\hat{\rho} = -2.53$, which means that $\hat{\lambda}_3 = \exp\{-2.53\} \approx 0.08$ in the bivariate Poisson fit.

As broadly explained in the previous sections, football models easily incorporate latent parameters describing the attacking and defensive skills of all the teams included in the dataset or for a selected portion. Once we fit a model with the footBayes package, the step of displaying and analyzing the teams' abilities is crucial and can be accomplished from both a graphical and numerical perspective. However, rather than printing a collection of numerical estimates, the function foot_abilities takes both "stanFoot" and "list" class objects and clearly depicts posterior/confidence intervals for the attacking and defensive abilities on the considered data (here for the bivariate Poisson only), as explained in Code Snippet 17.

Code Snippet 17 Italian Serie A 2009/2010: static abilities.

```
## plotting static abilities

foot_abilities(biv_stan, italy_2009)
foot_abilities(biv_mle, italy_2009)
```

In Output 13 we depict the 95% credibility and confidence intervals for the attacking (red segments) and defensive (blue segments) abilities for the twenty teams competing in the Italian Serie A 2009/2010: coherently with the model formulation in Section 4.4.2, the higher (lower) is the

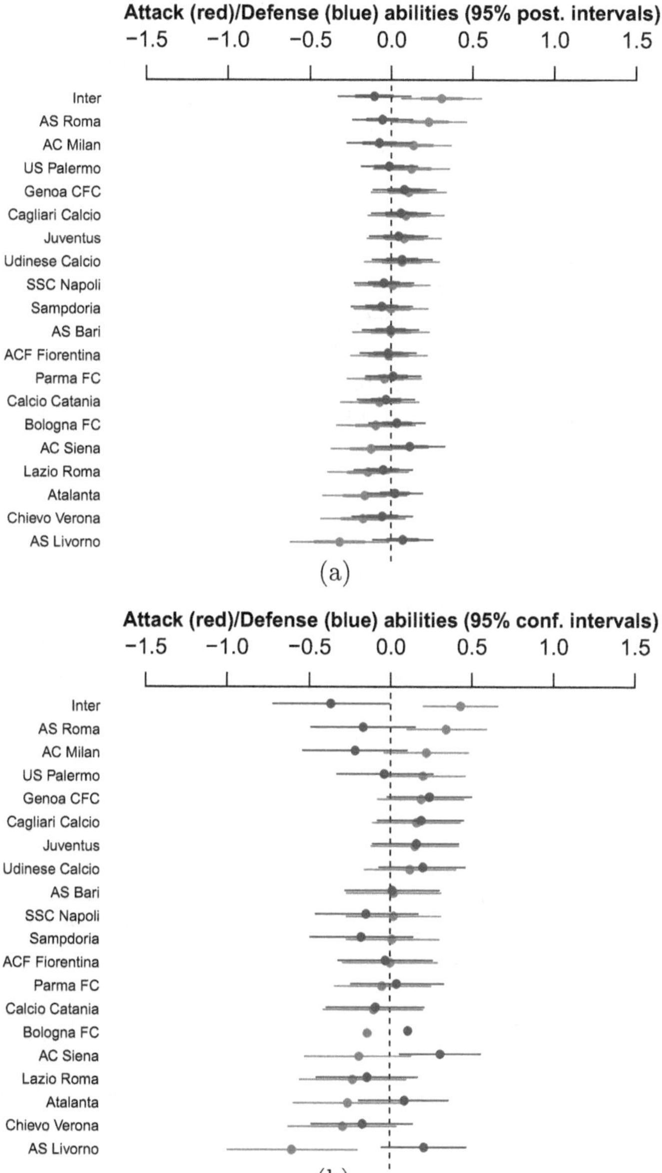

Output 13: Italian Serie A 2009/2010: 50% (thicker segments) and 95% (thinner segments) credible (a) and confidence (b) intervals for attacking (red segments) and defensive (blue segments) static abilities for the bivariate Poisson fit.

attack (defence) the better is evaluated the team. Internazionale FC (abbreviated with Inter), the team winning the Serie A 2009/2010, reports the best defensive and attacking skills; conversely, AS Livorno, which concluded the 2009/2010 season at the least ranking position and then has been relegated in the second division—the Italian Serie B—displays a relatively high defensive estimated value—again, representing weak defensive performance—and a low attacking estimated ability. The two plots are ordered according to the attacking ranks, with the best teams of Serie A that are correctly associated with high attacking and low defensive values: we remind again that these estimates are defined on the logarithmic scale, being included in the log-linear predictors for the average goals scored by two competing teams in each match.

In general, both the frequentist and the Bayesian model fits seem to well capture the static abilities, and yield similar results in terms of skills' estimates: in the next sections we will compare these estimated abilities with the final observed rank positions.

4.5.2 Dynamic models

As thoroughly motivated in Section 4.4.3, a structural limitation in the previous models is the assumption of static team-specific parameters, such that teams are assumed to show a constant performance across time, as determined by the attack and defence abilities. However, teams' performances tend to be dynamic and change across different years, if not different weeks when data related to domestic leagues, for many aforementioned factors, such as the players' transfermarket, injuries, dismissions, turnovers, etc.. For these and other many reasons, we can then assume dynamic abilities patterns using the dynamic_type argument in the stan_foot function, with possible options "seasonal" or "weekly" in order to consider more seasons or more week-times within a single season, respectively. We fit now the two double and the bivariate Poisson models by allowing weekly dynamic team-specific abilities along the Serie A 2009/2010 season and extract the posterior estimates for the fixed-effects parameters (Code Snippets 18 and 19), where the prior distributions for the offensive/defensive abilities are defined as in Equations (4.24) and (4.26).

Code Snippet 18 Italian Serie A 2009/2010: dynamic double Poisson model.

```
### Fit dynamic Bayesian models

dp_stan_dyn <- stan_foot(data = italy_2009,
                         model= "double_pois",
                         home_effect = TRUE,
                         dynamic_type ="weekly") # double poisson
print(dp_stan_dyn, pars =c("home", "sigma_att", "sigma_def"))
```

```
Summary of Stan football model
------------------------------

Posterior summaries for model parameters:
           mean se_mean   sd 2.5%  25%  50%  75% 97.5% n_eff Rhat
home       0.41       0 0.04 0.33 0.38 0.41 0.44  0.49  2062 1.00
sigma_att 0.04       0 0.01 0.02 0.03 0.04 0.04  0.06    24 1.09
sigma_def 0.03       0 0.01 0.02 0.02 0.03 0.04  0.05     9 1.42
```

Output 14: Dynamic double Poisson model's summary from stan_foot. Posterior estimates for the selected parameters.

Code Snippet 19 Italian Serie A 2009/2010: dynamic biv. Poisson model.

```
biv_stan_dyn <- stan_foot(data = italy_2009,
                          model= "biv_pois",
                          home_effect = TRUE,
                          dynamic_type ="weekly") # biv. Poisson
print(biv_stan_dyn, pars =c("home", "rho", "sigma_att", "sigma_def"))
```

```
Summary of Stan football model
------------------------------

Posterior summaries for model parameters:
           mean se_mean   sd  2.5%   25%   50%   75% 97.5% n_eff Rhat
home       0.33    0.00 0.05  0.23  0.30  0.34  0.37  0.43  2816 1.00
rho       -2.35    0.01 0.42 -3.34 -2.59 -2.29 -2.05 -1.66  3044 1.00
sigma_att  0.04    0.01 0.01  0.01  0.03  0.04  0.05  0.06     5 1.68
sigma_def  0.04    0.00 0.01  0.02  0.03  0.04  0.04  0.06    19 1.25
```

Output 15: Dynamic bivariate Poisson model's summary from stan_foot. Posterior estimates for the selected parameters.

The estimates for the home effect and the correlation parameter in the bivariate Poisson fit in Output 15 appear quite close to those previously obtained under the static model in Section 4.5.1, being $\hat{\lambda}_3 = \exp\{-2.35\} \approx 0.095$ and home $= 0.33$. According to the estimation of the attacking and defensive standard deviations (see Equation (4.24)), there is a slight lack of convergence of the algorithm under both the models as monitored by the Gelman-Rubin statistic, which is usually greater than the "golden" threshold 1.1: in this case, increasing the number of iterations (here by defaults there are 2000 HMC iterations) could help.

We can then depict the team-specific dynamic abilities, along with a 50% credible interval, by simply typing the following instructions reported in Code Snippet 20.

Code Snippet 20 Italian Serie A 2009/2010: dynamic abilities.

```
## Plotting dynamic abilities: credible 50% intervals

foot_abilities(biv_stan_dyn, italy_2009)
```

As we can see from Output 16, dynamic abilities naturally evolve over time: the best teams—Inter, AS Roma, AC Milan—report increasing attacking and decreasing defensive abilities, whereas the worst ones—AS Livorno, AC Siena and Atalanta—exhibit decreasing attacking and increasing defensive skills. The reason behind these increasing/decreasing trends is straightforward: the attack/defence parameters have been initialized by default to have a prior mean equal to zero, thus, as the season evolves and new information is acquired, the posterior estimates are computed, and these could oscillate much far away from the initial prior guess. We suggest the user to change the prior mean for the attacking/defensive parameters, by specifying his/her initial values to better reflect his/her feelings about the teams' strengths through the argument prior_par in the stan_foot function, as shown in the next section.

4.5.3 Changing default prior distributions

A common practice in Bayesian statistics is to change the prior distributions and perform some sensitivity checks with respect to the choice of some hyper-parameters. The default priors for the static team-specific abilities supported by the package are defined in Equation (4.7), whereas the default prior distributions for their related team-level standard deviations are given by Equation (4.9). For μ_{att} and μ_{def} the user could supply some fixed numerical values. However, the user is free to elicit some different priors for the abilities, choosing one among the following distributions: Gaussian (normal), student-t (student_t), Cauchy (cauchy) and Laplace (laplace). The prior_par optional argument allows to specify the priors for the team-specific parameters att and def through the argument ability, whereas a different prior for the team-specific standard deviations $\sigma_{\text{att}}, \sigma_{\text{def}}$ could be specified through the argument ability_sd. For instance we could consider:

$$\begin{aligned} att_k &\sim t_4(\mu_{\text{att}}, \sigma_{\text{att}}), \\ def_k &\sim t_4(\mu_{\text{def}}, \sigma_{\text{def}}), \; k = 1, 2, , \ldots, K \\ \sigma_{\text{att}}, \sigma_{\text{def}} &\sim \text{Laplace}^+(0, 1), \end{aligned} \tag{4.32}$$

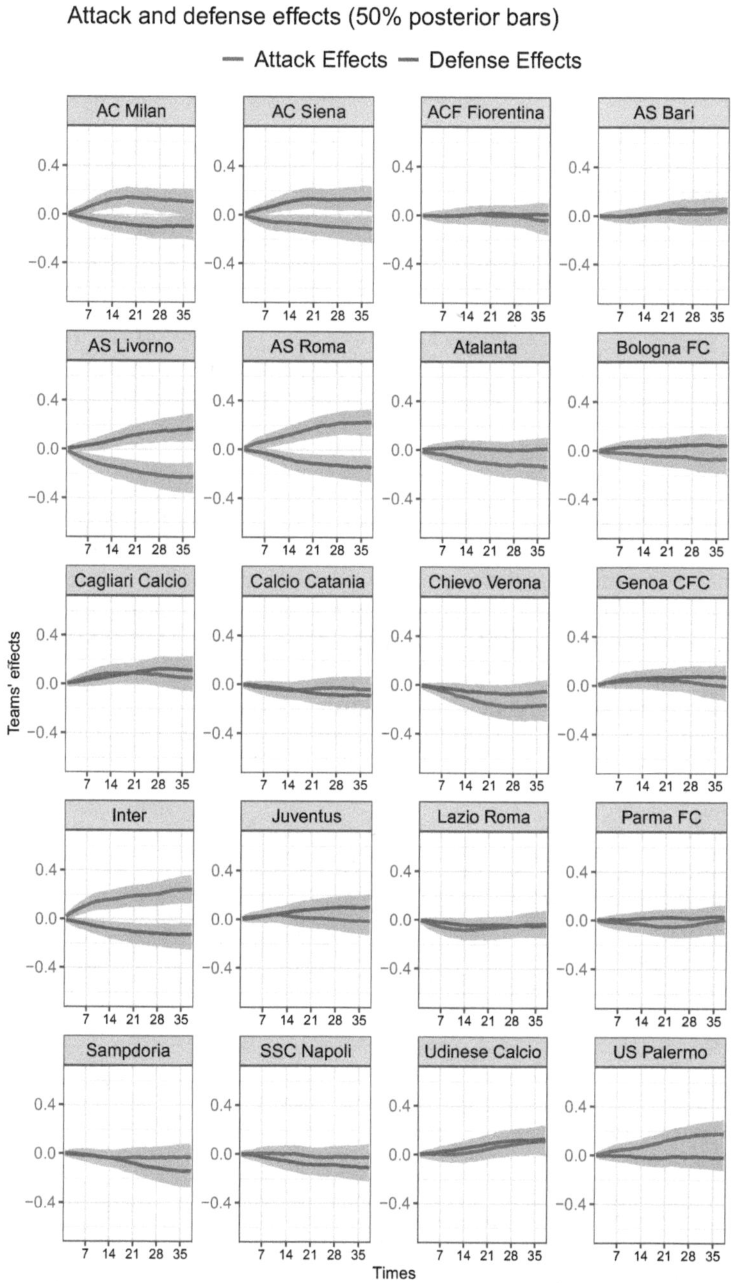

Output 16: Italian Serie A 2009/2010: 50% credible intervals (grey ribbons) for attack (red curves) and defence (blue curves) weekly dynamic abilities for the dynamic bivariate Poisson fit.

where $t_{\text{df}}(\mu, \sigma)$ denotes a student-t distribution with df degrees of freedom, location μ and scale σ, whereas Laplace^{+} denotes a half-Laplace distribution. Code Snippet 21 shows how to change the prior distributions accordingly.

Code Snippet 21 Italian Serie A 2009/2010: change prior distributions.

```
## change default priors
biv_stan_t <- stan_foot(data = italy_2009,
                        model="biv_pois",
                        home_effect = TRUE,
                        chains = 4,
                        prior_par = list(ability = student_t(4,0,NULL),
                                         ability_sd = laplace(0,1)),
                        iter = n_iter) # biv poisson
```

Then, we can compare the marginal posteriors from the two models, the bivariate Poisson with default Gaussian team-specific abilities and the Cauchy$(0, 5)$ prior for the team-level standard deviations, and the other one specified above, with student-t distributed team-specific abilities and the Laplace$^{+}(0, 1)$ prior for the team-level standard deviations: the code is reported in Code Snippet 22. We depict in Output 17 the comparison for the attacking team-specific standard deviations only, by combining the bayesplot function mcmc_areas with some optional ggplot2 arguments as reported in Code Snippet 22.

Code Snippet 22 Italian Serie A 2009/2010: marginal posterior comparison.

```
## graphical posteriors comparison

posterior1_t <- as.matrix(biv_stan_t$fit)
model_names <- c("Gauss. + Cauchy", "Stud+Laplace")
color_scheme_set(scheme = "gray")
gl_posterior <- cbind(posterior1[,"sigma_att"],
                      posterior1_t[,"sigma_att"])
colnames(gl_posterior)<-c("sigma_att", "sigma_att_t")
mcmc_areas(gl_posterior, pars=c("sigma_att", "sigma_att_t"))+
  xaxis_text(on =TRUE, size=ggplot2::rel(2.9))+
  yaxis_text(on =TRUE, size=ggplot2::rel(2.9))+
  scale_y_discrete(labels = ((parse(text= model_names))))+
  ggtitle(expression(sigma[att]))+
  theme(plot.title = element_text(hjust = 0.5, size =rel(2.6)))
```

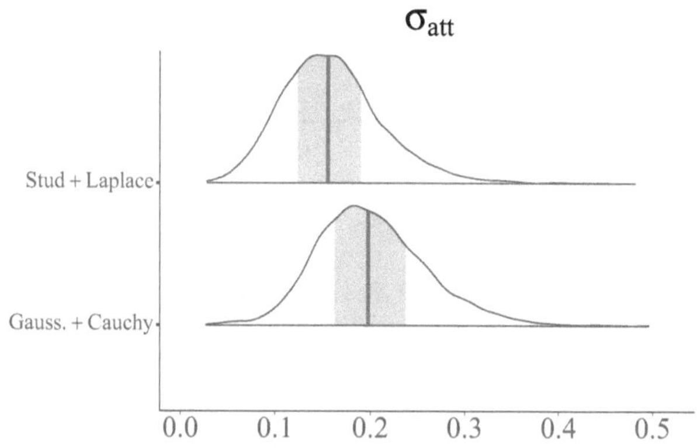

Output 17: Italian Serie A 2009/2010: marginal posterior distributions for the attacking team-specific standard deviation σ_{att} from: the bivariate Poisson model with student-t and half-Laplace priors; the bivariate Poisson model with default Gaussian and half-Cauchy priors. The grey area denotes the 50% high posterior density (HPD) interval.

The half-Laplace prior induces a slightly lower amount of group-variability, then, a slightly larger amount of shrinkage towards the grand mean μ_{att}.

When specifying the prior for the team-specific parameters through the argument prior_par, the user is not allowed to supply the group-level standard deviations $\sigma_{\text{att}}, \sigma_{\text{def}}$ with some numerical values. Rather, they need to be assigned a reasonable prior distribution. For such reason, the most appropriate specification for the ability argument is through the following pseudo-code: prior_par = list(ability = "dist"(0, NULL)), where the standard deviation/scale argument is set to NULL—otherwise, a warning message occurs.

4.5.4 Predictions and predictive accuracy

4.5.4.1 Posterior out-of-sample probabilities

The hottest feature in sports analytics is to obtain predictions for future matches. We consider the posterior predictive distribution—hereafter, ppd—for future and observable data $\tilde{\mathcal{D}}$ and acknowledge then the whole model's prediction uncertainty, which propagates from the posterior model's uncertainty. In such a way, we can generate observable values $\tilde{D}$ conditioned on the data $\mathcal{D} = (y_{i1}, y_{i2})_{i=1,\ldots,n}$:

$$f(\tilde{\mathcal{D}}|\mathcal{D}) = \int f(\tilde{\mathcal{D}}|\boldsymbol{\theta}) f(\boldsymbol{\theta}|\mathcal{D}) d\boldsymbol{\theta},$$

where $f(\boldsymbol{\theta}|\mathcal{D})$ is the posterior distribution of the parameter vector $\boldsymbol{\theta}$. We may then predict held-out match results by using the argument predict of the stan_foot function, considering the last two weeks of the 2009/2010 season as the test set. The function foot_prob allows to compute posterior-results probabilities for the selected matches, in terms of home win, draw, and away win probabilities (see Code Snippet 23).

Code Snippet 23 Italian Serie A 2009/2010: probabilistic predictions.

```
## predict the last 2 match-days of the season

biv_stan_pred <- stan_foot(data = italy_2009,
                           model="biv_pois",
                           home_effect = TRUE,
                           predict = 20,
                           iter = n_iter)  # biv poisson
foot_prob(object = biv_stan_pred, data = italy_2009)
```

Table 4.1 returned by the foot_prob function reports the home and the away teams in the first two columns. Columns from three to five report the

TABLE 4.1
Italian Serie A 2009/2010: probabilistic predictions for held-out matches, the last two match-days under the static bivariate Poisson model. Columns from three to five report the home win, draw, and away win posterior predictive probabilities, respectively, whereas the sixth column reports the "most likely result" (MLO) according to the collection of all the possible results as returned by the ppd

Home team	Away team	Home win	Draw	Away win	MLO
SSC Napoli	Atalanta	0.518	0.257	0.225	1–0 (0.126)
Udinese Calcio	AS Bari	0.467	0.255	0.278	1–1 (0.109)
AS Roma	Cagliari Calcio	0.535	0.230	0.235	1–1 (0.098)
Bologna FC	Calcio Catania	0.433	0.274	0.293	1–0 (0.12)
Inter	Chievo Verona	0.587	0.231	0.182	1–0 (0.117)
AS Livorno	Lazio Roma	0.370	0.305	0.325	1–0 (0.141)
Genoa CFC	AC Milan	0.419	0.250	0.331	1–1 (0.109)
Juventus	Parma FC	0.504	0.252	0.244	1–1 (0.115)
US Palermo	Sampdoria	0.489	0.251	0.260	1–1 (0.118)
ACF Fiorentina	AC Siena	0.524	0.249	0.227	1–1 (0.111)
AC Milan	Juventus	0.505	0.245	0.250	1–1 (0.108)
Lazio Roma	Udinese Calcio	0.415	0.276	0.309	1–1 (0.128)
Cagliari Calcio	Bologna FC	0.519	0.243	0.238	1–1 (0.111)
AS Bari	ACF Fiorentina	0.430	0.268	0.302	1–1 (0.127)
Calcio Catania	Genoa CFC	0.436	0.256	0.308	1–1 (0.117)
AC Siena	Inter	0.299	0.250	0.451	1–1 (0.107)
Parma FC	AS Livorno	0.509	0.273	0.218	1–0 (0.153)
Sampdoria	SSC Napoli	0.451	0.266	0.283	1–1 (0.121)
Atalanta	US Palermo	0.376	0.271	0.353	1–1 (0.131)
Chievo Verona	AS Roma	0.340	0.272	0.388	1–1 (0.126)

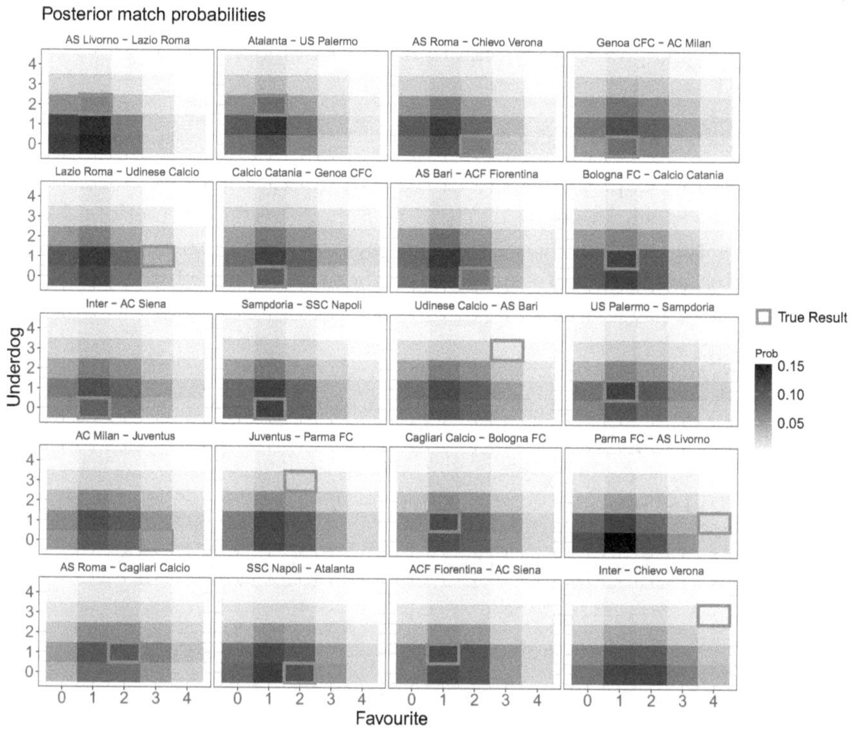

Output 18: Italian Serie A 2009/2010: posterior predictive probabilities results for the last two match-days under the static bivariate Poisson model. Darker regions correspond to results with higher probabilities, whereas the red square denotes the final actual result.

home win, draw, and away win posterior predictive probabilities, respectively, whereas the sixth column reports the "most likely result" (MLO) according to the collection of all the possible results as returned by the ppd.

The function foot_prob returns also some "*chessboard*" plots for the held-out matches as those provided in Output 18, where darker regions are associated with higher posterior probabilities, whereas the red square corresponds to the actual final observed results. As a matter of rough interpretation, when the red square is in correspondence of darker regions, this denotes an approximated good agreement between the prediction and the actual result. In this plot the first team name in the single labels denotes the favourite team, whereas the second name denotes the "underdog" team where the term "underdog" refers to the team associated with lower winning chances; moreover, we depict the most balanced matches to the most apparently unbalanced ones from the left top corner to the right bottom corner. Thus, the match AS Livorno-Lazio Roma (top left corner in the plot) is the most balanced, whereas Inter-Chievo Verona (right bottom corner) the most unbalanced in favour of Inter, whose the predicted winning probability is about 59%.

4.5.5 Rank-league reconstruction

As widely explained in Chapter 3, Sections 3.3 and 3.3.2, statisticians and football amateurs are much interested in the final rank-league predictions. However, predicting the final rank positions, along with the corresponding teams' points, is often assimilated to an oracle, rather than a coherent statistical procedure. We can provide here:

- *retrospective rank-league reconstruction* (aggregated or at team-level) by using the foot_rank function and in-sample replications $\mathcal{D}^{\text{rep}}$, as depicted in Outputs 19-20 for the bivariate Poisson model, where yellow ribbons denote the credible intervals, and solid blue points and lines denote the observed final and cumulated points, respectively (Code Snippet 24).

Code Snippet 24 Italian Serie A 2009/2010: rank-league reconstruction.

```
## Retrospective rank reconstruction
# aggregated plot

foot_rank(data = italy_2009, object = biv_stan)

# team-specific plot

foot_rank(data = italy_2009, object = biv_stan,
          visualize = "individual")
```

- *Ahead rank-league prediction* (aggregated or at team-level) by using the foot_rank function and the held-out replications $\tilde{\mathcal{D}}$ as depicted in Table 4.2 and Outputs 21-22 for the bivariate Poisson model, where yellow ribbons denote the credible intervals, and solid blue points and lines denote the observed final and cumulated points, respectively (Code Snippet 25). Training set: first 36 match-days; test set: last two match-days.

  In Table 4.2, the third column reports for each team the predicted number of points in terms of medians, the fourth column indicates the number of observed points, whereas the fifth column contains the 50% credibility interval from ppd for the predicted points under the model. As we may easily notice, the model predicted points match pretty well the observed points for each team.

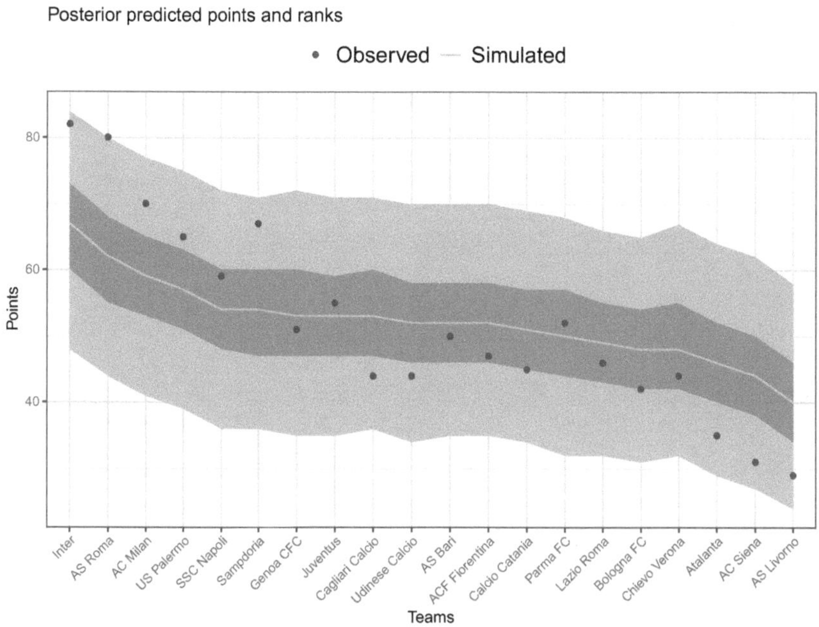

Output 19: Italian Serie A 2009/2010: retrospective rank-league reconstruction using in-sample replications from the posterior predictive distribution under the static bivariate Poisson model. The blue dots denote the final observed points; dark yellow ribbons denote 50% credible intervals, and light yellow ribbons denote 95% credible intervals from the ppd.

4.5.6 Model checking

Checking the model fit is a relevant and vital statistical task. To this purpose, we can evaluate hypothetical replications $\mathcal{D}^{\text{rep}}$ under the posterior predictive distribution:

$$f(\mathcal{D}^{\text{rep}}|\mathcal{D}) = \int f(\mathcal{D}^{\text{rep}}|\boldsymbol{\theta})f(\boldsymbol{\theta}|\mathcal{D})d\boldsymbol{\theta},$$

and check whether these replicated values are somehow close to the observed data $\mathcal{D}$: in other words, we want to check whether the observed values are a plausible realization of the simulation process obtained through the ppd. These methods aimed at comparing hypothetical replications with the observed data are named *posterior predictive checks* (Gelman et al., 1996, 2013) and nowadays they represent a milestone for goodness-of-fit purposes in modern Bayesian inference.

The function pp_foot allows to obtain:

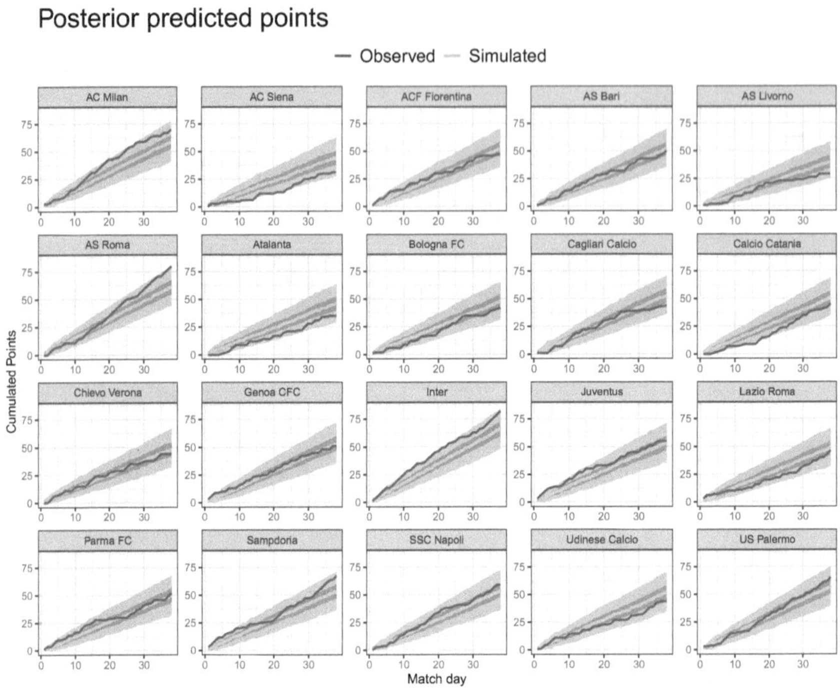

Output 20: Italian Serie A 2009/2010: retrospective rank-league reconstruction using in-sample replications from the posterior predictive distribution under the static bivariate Poisson model for each team. The blue lines denote the cumulated observed points; dark yellow ribbons denote 50% credible intervals, and light yellow ribbons denote 95% credible intervals from the ppd for the simulated cumulated points.

- an aggregated plot depicting the frequencies of the observed goal differences $Z_i = Y_{i1} - Y_{i2}$, $i = 1, \dots, n$ in the dataset (blue segments) plotted against the replicated ones (orange jittered points)—those obtained through the ppd—as depicted in Output 24 if the argument `type = "aggregated"` is chosen. In this case the function returns also in the R console the Bayesian p-value (Gelman et al., 1996), $\Pr(Z_i^{\text{rep}} \geq Z_i | \mathcal{D})$, to monitor any possible misfit, as shown in Output 23—note that values for these p-values too close to 0 or 1 are symptomatic of model deficiencies—see Code Snippet 26 for the corresponding implementation.

  NOTE: goal differences greater than three in absolute value are not considered here.

- A visualization of the match-ordered goal differences (blue segments) along with their 50% and 95% credible intervals (light and dark orange ribbons) obtained from the ppd as depicted in Output 26 if the argument `type =`

Code Snippet 25 Italian Serie A 2009/2010: rank-league predictions.

```
## Rank predictions

# aggregated plot

foot_rank(data = italy_2009, object = biv_stan_pred)

# team-specific plot

foot_rank(data = italy_2009, object = biv_stan_pred,
          visualize = "individual",
          teams = c( "Inter", "AS Roma",
                       "AC Milan", "Sampdoria"))
```

"matches" is chosen. In this case the function returns in the R console a short table displaying the level of credibility chosen by the user—coverage = 0.95 is the default option—and a measure of empirical coverage to check

TABLE 4.2
Italian Serie A 2009/2010: rank-league predictions under the static bivariate Poisson model. Training data: first 36 match-days; test data: last two match-days. The third column reports the posterior predicted medians for the final points, whereas the fifth column reports the 50% credible interval for the final points

Position	Team	Estimated (50%)	Observed	50% Interval
1	Inter	80	82	(79, 82)
2	AS Roma	77	80	(76, 78)
3	AC Milan	70	70	(68, 71)
4	Sampdoria	66	67	(64, 67)
5	US Palermo	64	65	(62, 65)
6	SSC Napoli	59	59	(57, 60)
7	Juventus	58	55	(56, 59)
8	Genoa CFC	51	51	(49, 52)
9	ACF Fiorentina	49	47	(47, 50)
10	AS Bari	49	50	(47, 50)
11	Parma FC	49	52	(47, 50)
12	Udinese Calcio	46	44	(44, 47)
13	Cagliari Calcio	46	44	(44, 47)
14	Chievo Verona	46	44	(45, 47)
15	Calcio Catania	44	45	(42, 45)
16	Bologna FC	43	42	(41, 44)
17	Lazio Roma	43	46	(41, 44)
18	Atalanta	38	35	(36, 38)
19	AS Livorno	32	29	(30, 32)
20	AC Siena	32	31	(30, 33)

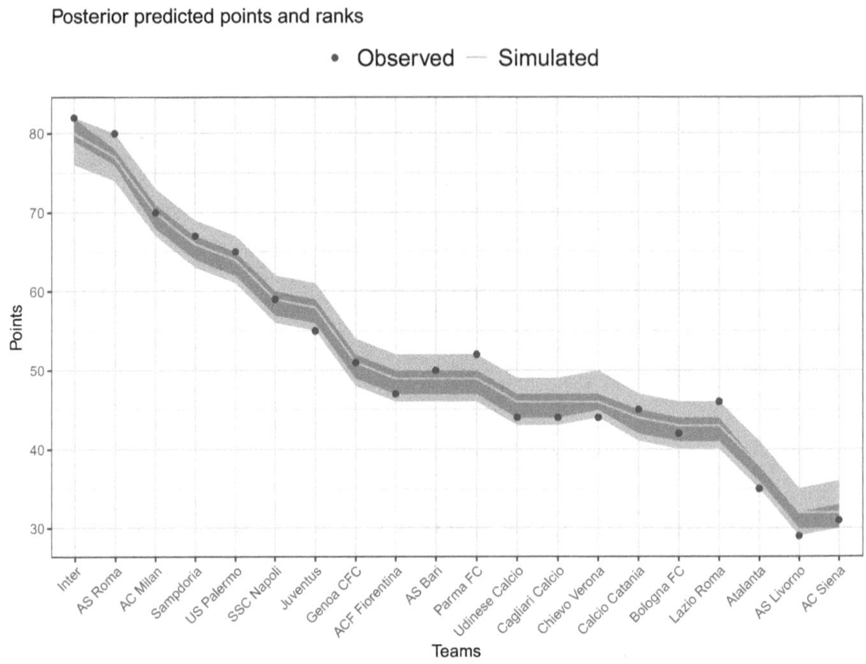

Output 21: Italian Serie A 2009/2010: rank-league prediction using out-of-sample replications from the posterior predictive distribution under the static bivariate Poisson model (training set: first 36 match-days; held-out set: last two match-days). The blue dots denote the final observed points; dark yellow ribbons denote 50% credible intervals, and light yellow ribbons denote 95% credible intervals from the ppd.

whether the observed points fall in the corresponding credible intervals—note that values for the empirical coverage much smaller than the credibility level could be symptomatic of model deficiencies, conversely values higher than the credibility level could suggest an amount of overfitting (see Output 25 and Code Snippet 27).

Code Snippet 26 Italian Serie A 2009/2010: aggregated model checking.

```
## PP checks: aggregated goal's differences
   and ordered goal differences

pp_foot(data = italy_2009, object = biv_stan,
        type = "aggregated")
```

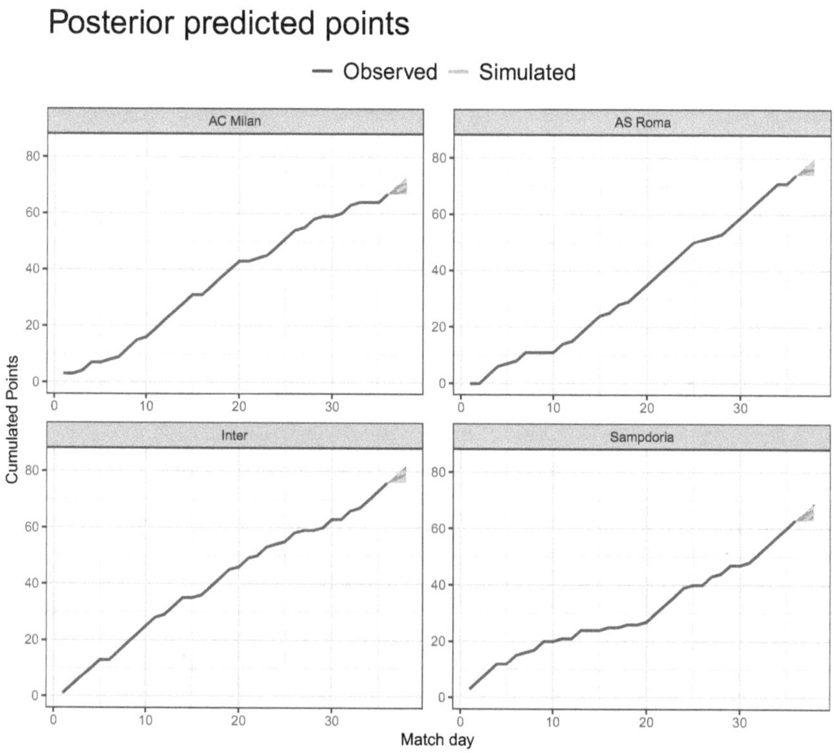

Output 22: Italian Serie A 2009/2010: rank league prediction using out-of-sample replications from the posterior predictive distribution for the four best teams in that season, namely Inter, AS Roma, AC Milan and Sampdoria under the static bivariate Poisson model (training-set: 36 first match-days; held-out set: last two match-days). The blue lines denote the cumulated observed points; dark yellow ribbons denote 50% credible intervals, and light yellow ribbons denote 95% credible intervals from the ppd for the simulated cumulated points.

```
$pp_table
  goal diff. Bayesian p-value
1         -3            1.000
2         -2            0.758
3         -1            0.667
4          0            0.227
5          1            0.174
6          2            0.150
7          3            0.300
```

Output 23: Italian Serie A 2009/2010: model checking table for the observed goal differences under the static bivariate Poisson model. The first column indicates the goal differences from -3 until 3, and the second column reports the Bayesian p-value for each goal-difference.

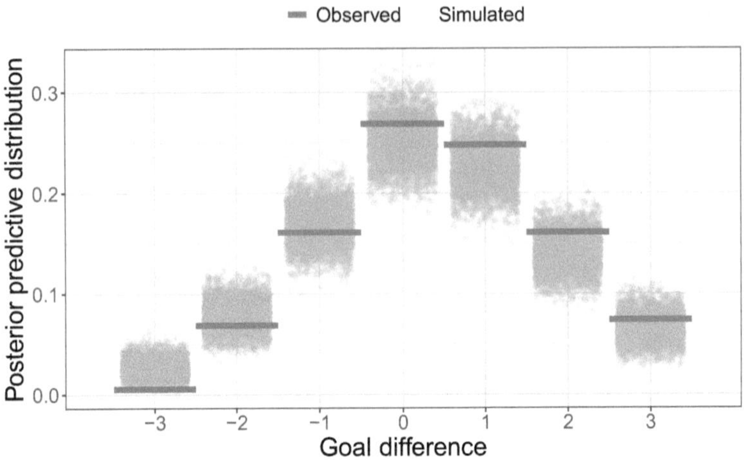

Output 24: Italian Serie A 2009/2010: frequencies of observed goal differences (blue segments) against replicated goal differences from the ppd (orange jittered points) under the static bivariate Poisson model.

Code Snippet 27 Italian Serie A 2009/2010: matches' model checking.

```
pp_foot(data = italy_2009, object = biv_stan,
        type = "matches")
```

```
$pp_table
  1-alpha emp. coverage
1    0.95         0.968
```

Output 25: Italian Serie A 2009/2010: model checking table for single matches.

The aggregated goal difference frequencies seem to be decently captured by the model's ppd replications: in the first plot, in Output 24, the blue horizontal lines denote the observed goal differences frequencies registered in the dataset, whereas the yellow jittered points denote the corresponding ppd replications. A goal-difference of 0, corresponding to the draws, is not underestimated by the model—for draw inflation arguments, see models in Chapter 5—and the corresponding p-value is in fact equal to 0.227. The only problematic case occurs for a goal-difference of −3, which corresponds to an away team beating an home team with three goals of margin: in this case, the static BP model overestimates this occurrence, and the Bayesian p-value is equal to 1, being all the orange jittered points greater than the observed frequency.

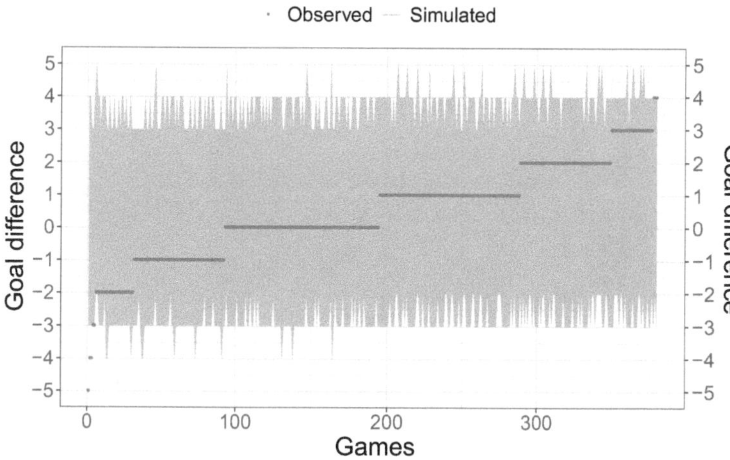

Output 26: Italian Serie A 2009/2010: ordered observed goal differences (blue segments) against 95% ordered credible intervals for simulated the goal differences (orange ribbons) obtained from the ppd under the static bivariate Poisson model.

In the second plot, in Output 26, the ordered observed goal differences (blue segments) are plotted against their corresponding 50% (dark orange ribbons) and 95% (light yellow ribbons) credible intervals as obtained from the ppd: also from this plot we do not notice particular signs of model's misfits, and this is also suggested by the empirical coverage in Output 25, equal to 0.968.

Other useful PP checks designed to reveal model's inconsistencies, such as the overlap between data density estimation and replicated data densities estimations obtained from the ppd, can be obtained through the standard use of the bayesplot package, for instance providing an approximation to a continuous distribution using an input kernel choice (bw = 0.5 in the ppc_dens_overlay function) as reported in Code Snippet 28.

Code Snippet 28 Italian Serie A 2009/2010: goal-difference.

```
## PPC densities overlay with the bayesplot package
# extracting the replications

sims <-rstan::extract(biv_stan$fit)
goal_diff <- italy_2009$home_goals-italy_2009$away_goals

# plotting data density vs replications densities
ppc_dens_overlay(goal_diff, sims$y_rep[,,1]-sims$y_rep[,,2], bw = 0.5)
```

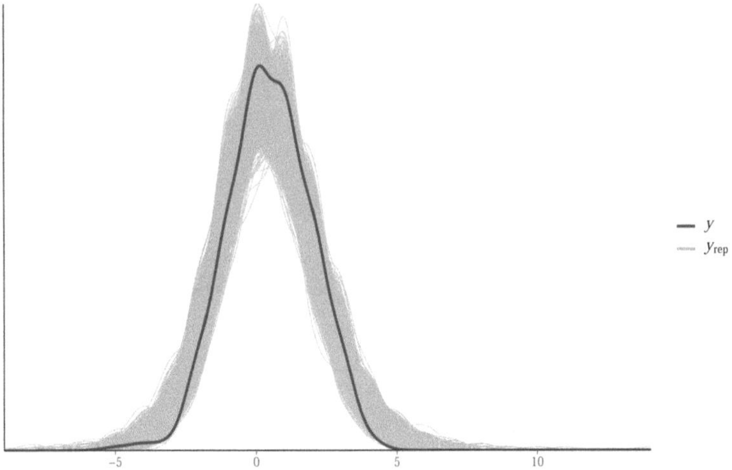

Output 27: Italian Serie A 2009/2010: density estimation for the observed goal difference (dark grey line) against the distribution of the goal difference (light grey lines) from the ppd of the static bivariate Poisson model.

From the plot reported in Output 27 above we get the empirical confirmation that the goal difference is well captured by the static bivariate Poisson model.

4.5.7 Model comparison with the loo package

Comparing statistical models in terms of some *predictive information criteria* should be conclusive in terms of model selection and may be carried out by using the leave-one-out cross-validation criterion (LOOIC) and the Watanabe Akaike Information criterion (WAIC)—see Chapter 2, Sections 2.9.3.2 and 2.9.3.3—performed by using the loo package. For more details about LOOIC and WAIC and other information criteria, see Gelman et al. (2014); Vehtari et al. (2017).

The general formulation for the predictive information criteria, such as the classical AIC and BIC, is the following:

$$\text{crit} = -2\widehat{\text{elpd}} = -2(\widehat{\text{lpd}} - \text{parameters penalty}),$$

where:

- $\widehat{\text{elpd}}$ is the estimate of the expected log predictive density of the fitted model;
- $\widehat{\text{lpd}}$ is a measure of the log predictive density of the fitted model;
- parameters penalty is a penalization accounting for the effective number of parameters of the fitted model.

The canonical interpretation is that the lower the information criterion, and the better is the estimated model's predictive accuracy. Moreover, if two competing models share the same value for the log predictive density, the model with less parameters is favoured, due to the well-known *Occam's rasor*.

We can perform some Bayesian model comparisons by using the loo and waic functions of the loo package. We compare the vanilla double Poisson model, the two static bivariate Poisson models—the one with default priors and the other with student-t and half-Laplace priors proposed in Section 4.5.3—and the weekly dynamic bivariate Poisson model in Section 4.5.2 fitted on the Italian Serie A 2009/2010, as reported in Code Snippet 29.

Code Snippet 29 Italian Serie A 2009/2010: model comparisons.

```
### Model comparisons
## LOOIC, loo function

# extract pointwise log-likelihood
log_lik_dp <- extract_log_lik(dp_stan$fit)             # static dp
log_lik_dp_dyn <- extract_log_lik(dp_stan_dyn$fit)     # dynamic dp
log_lik_biv <- extract_log_lik(biv_stan$fit)           # static biv-pois
log_lik_biv_t <- extract_log_lik(biv_stan_t$fit)       # static biv pois t
log_lik_biv_dyn <- extract_log_lik(biv_stan_dyn$fit)  # dynamic biv-pois

# compute loo
loo_dp <- loo(log_lik_dp)
loo_biv <- loo(log_lik_biv)
loo_biv_t <- loo(log_lik_biv_t)
loo_dp_dyn <- loo(log_lik_dp_dyn)
loo_biv_dyn <- loo(log_lik_biv_dyn)


# compare five looic
compare(loo_dp, loo_biv,
        loo_biv_t, loo_dp_dyn, loo_biv_dyn)
```

TABLE 4.3
Italian Serie A 2009/2010: model comparisons via LOOIC. The second column reports the effective number of parameters as estimated through the procedure

Model	Eff. parameters	LOOIC
Vanilla double Poisson	18.6	2138.7
Biv. Poisson (student-t priors)	16.7	2139.7
Biv. Poisson (default priors)	17.4	2139.9
Dynamic double Poisson	22.0	2143.2
Dynamic bivariate Poisson	22.7	2145.8

According to the above LOOIC comparisons in Table 4.3, where the second column reports the effective number of parameters estimated for each model and the third column indicates the corresponding LOOIC value, the static vanilla double Poisson model attains the lowest LOOIC value (2138.7) and is then the best model in terms of predictive accuracy; the bivariate Poisson with student-t and half-Laplace prior reports a LOOIC of 2139.7, whereas the weekly dynamic models report higher LOOICs, 2143.2 for the double Poisson and 2145.8 for the bivariate Poisson, respectively. Regarding the effective number of parameters, this value in the static models is approximately equal to 17 or 18, lower than the dynamic models, approximately equal to 22. For more details about these values, we invite the interested reader to consult the work of Vehtari et al. (2017) for more and deep details about LOOIC and some related measures. Despite the supremacy of the static models in this dataset, it is likely that in other datasets/seasons the assumption of static team-specific parameters and goal independence could result to be too restrictive and oversimplified to capture teams' skills over time and get the best predictive performance in terms of LOOIC/WAIC. Another aspect related to the given dataset is that the Italian Serie A usually yields an extent of under-dispersion in the scores data due to a defensive style of playing, and this happens also for the 2009/2010 season: the average home goal per match is 1.54, with a variance of 1.34, whereas the average away goal per match is 1.07 with a standard deviation of 0.99—we remind that the average number of goals and its variance are the same according to the Poisson distribution.

Anyway, combining these predictive comparisons with the model checking procedures proposed in Section 4.5.6 for the static models, it is evident how static team-specific models capture well the nature of the data; more in specific, the vanilla double Poisson model, as largely documented in previous chapters, is pretty adequate and may be used as a relevant milestone for these kinds of data, since it implies small computational costs and gives overall good fitting results.

4.6 Summary and closing remarks of Chapter 4

Fitting a goal-based model in R either with classical or Bayesian methods is an easy task. Unless one wants to write its own model, the package footBayes represents a unique tool that provides some simple and automatic functions to estimate a model, graphically display the estimates, check the model, and make out-of-sample predictions.

Section 4.1 explains how to install the package from CRAN, whereas the list of available basic models is detailed in Section 4.2, perhaps the user can choose one among the following models—some of them will be covered in Chapter 5—: double Poisson, bivariate Poisson, diagonal-inflated bivariate Poisson, Skellam, zero-inflated Skellam, and student-t. The basic syntax of the main functions is thoroughly provided in Section 4.3. Double Poisson, bivariate

Poisson and dynamic extensions are proposed in Sections 4.4.1, 4.4.2, and 4.4.3.

Section 4.5 presents a long case-study using the Italian Serie A 2009/2010 data. Double and bivariate static Poisson models are fitted in Section 4.5.1, and the team-specific abilities are therein plotted. A dynamic modelling extension is proposed in Section 4.5.2, whereas the change of the default prior distributions is the core of Section 4.5.3. Out-of-sample predictions obtained through the posterior predictive distributions are introduced in Section 4.5.4, whereas the retrospective reconstruction and the prediction of the final rank league table are provided in Section 4.5.5. Some Bayesian goodness-of-fit measures are detailed in Section 4.5.6, whereas Section 4.5.7 compares the basic models in terms of the leave-one-out cross-validation information criterion (LOOIC).

5

Additional statistical models for the scores

The art of posing a statistical model for a football match means specifying a random mechanism governing the final observed result. According to the existing football literature, and as broadly remarked in the previous chapters, the task may be accomplished via a *goal-based* approach for the number of goals, or via a *result-based* approach for the final result—home win, draw, away win. Although there is not a clear supremacy of one approach over the other, in this book we propose the main models formulated under the first framework, given their huge relevance in the scientific literature about football. To this aim, we proposed the two basic goal-based models, the double and the bivariate Poisson, in Chapter 4, and we showed how to fit them according to both maximum likelihood and MCMC methods.

We still denote with (Y_{i1}, Y_{i2}) the random variables representing the number of goals scored by the home and the away team in the i-th game, $i \in \{1, \dots, n\}$, respectively, where n is the total number of games—or matches, observations; the pair (y_{i1}, y_{i2}) denotes instead the observed number of goals for game i. The purpose of this Chapter is to introduce some more additional models accounting for draw inflation, goal correlation, and the inclusion of further covariates such as the ranking measures.

5.1 Other models available in footBayes

5.1.1 Diagonal-inflated bivariate Poisson

One of the main concerns in the implementation of double Poisson and bivariate Poisson models regards the so-called *draw inflation*, namely the natural tendency of these models to underestimate the observed number of draws arising in football data, as broadly anticipated in Section 2.8.3 of Chapter 2. As illustrated later in this Chapter in Section 5.4.1, the scaled double Poisson model (5.12) proposed by Dixon and Coles (1997) in the previous section is an inflated model attempt which actually inflates some occurrences (0-0, 1-0, 0-1, 1-1), including also some low scoring draws such as 0-0 and 1-1, by adding a

DOI: 10.1201/9781003186496-5

further parameter. Nonetheless, one may want to deal with more general models accounting for scores' dependence and able to improve the fit on the counts of draws. Following this intuition, Karlis and Ntzoufras (2003) propose a general modelling formulation which inflates the probabilities of draws, where a draw between two teams is represented by the outcomes on the diagonal of the probability table. To correct for the excess of draws one could add an inflation component on the diagonal of the probability function: the resulting model is an extension of the simple zero-inflated model that allows for an excess in 0-0 draws (Li et al., 1999). Then, an inflation of the draws probabilities designed to capture the draw occurrences could yield better models and allows for overdispersed, relative to the simple Poisson distribution, marginal distributions.

We consider the bivariate Poisson (4.16) in Chapter 4, Section 4.4.2, as the starting model, a diagonal-inflated model is then specified by the following probabilistic law:

$$f_{Y_1,Y_2}(y_1,y_2) = \begin{cases} (1-p)\text{BP}(y_1,y_2|\lambda_1,\lambda_2,\lambda_3), & y_1 \neq y_2 \\ (1-p)\text{BP}(y_1,y_2|\lambda_1,\lambda_2,\lambda_3) + pD(y_1,\theta), & y_1 = y_2, \end{cases} \tag{5.1}$$

where $D(y_1,\theta)$ is a discrete distribution with parameter vector θ and p is the probability of drawn inflation; useful choices for $D(y_1,\theta)$ are the geometric, the Poisson, or even the Bernoulli distribution. Such models can be fitted by using the EM algorithm algorithm—see Section 2.3.2 in Chapter 2.

Although univariate zero-inflated Poisson regression models have been developed and examined in detail (Lambert, 1992; Böhning et al., 1999), multivariate extensions, similar to the models proposed in this paper, are relatively rare with the exception of Li et al. (1999); Gan (2000); Walhin (2001).

There are two important properties of such models. Firstly, the marginal distributions of a diagonal-inflated model are not Poisson distributions but mixtures of distributions with one Poisson component. Secondly, if $\lambda_3 = 0$—corresponding to the double Poisson distribution (4.1)—the resulting inflated distribution introduces a degree of dependence between the two variables under consideration. For this reason, diagonal inflation may correct both the overdispersion and the correlation problems that are usually encountered in modelling football games and, as claimed by Karlis and Ntzoufras (2003), these models usually provide a better fit on football data.

5.1.1.1 Implementation in footBayes

In footBayes the user can fit a diagonal-inflated bivariate Poisson model, via HMC estimation, through the following code in Code Snippet 30 using the stan_foot function, whose summary is reported in Output 28.

Code Snippet 30 Italian Serie A 2009/2010: DIBP model.

```
###  Other models
## Diagonal-inflated bivariate Poisson

dibiv_stan <- stan_foot(data = italy_2009,
                        model= "diag_infl_biv_pois",
                        home_effect = TRUE)
print(dibiv_stan, pars = c("home", "rho", "sigma_att",
                           "sigma_def", "prob_of_draws"))
```

```
Summary of Stan football model
------------------------------

Posterior summaries for model parameters:
               mean se_mean   sd  2.5%   25%   50%   75% 97.5% n_eff Rhat
home           0.42    0.00 0.05  0.32  0.38  0.42  0.45  0.51  4383    1
rho           -4.22    0.01 0.45 -5.15 -4.52 -4.19 -3.89 -3.42  5810    1
sigma_att      0.21    0.00 0.06  0.11  0.17  0.20  0.25  0.35  1240    1
sigma_def      0.12    0.00 0.06  0.02  0.07  0.11  0.15  0.24   302    1
prob_of_draws 0.24    0.00 0.02  0.20  0.23  0.24  0.26  0.29  5391    1
```

Output 28: Italian Serie A 2009/2010: model's summary of the diagonal inflated bivariate Poisson from stan_foot. Posterior estimates for the selected parameters.

A dynamic version of the model allowing for weekly dynamic team-specific abilities can be estimated through the same function as shown in Code Snippet 31. The model's summary is reported in Output 29.

Code Snippet 31 Italian Serie A 2009/2010: dynamic DIBP model.

```
dibiv_stan_dyn <- stan_foot(data = italy_2009,
                            model= "diag_infl_biv_pois",
                            home_effect = TRUE,
                            dynamic_type = "weekly")
print(dibiv_stan_dyn, pars = c("home", "rho", "sigma_att",
                               "sigma_def", "prob_of_draws"))
```

```
Summary of Stan football model
------------------------------

Posterior summaries for model parameters:
              mean se_mean   sd  2.5%   25%   50%   75% 97.5% n_eff Rhat
home          0.41    0.00 0.05  0.31  0.38  0.41  0.45  0.51  3084 1.01
rho          -4.22    0.01 0.46 -5.24 -4.51 -4.20 -3.90 -3.40  3527 1.00
sigma_att     0.04    0.00 0.01  0.02  0.03  0.04  0.05  0.07    11 1.39
sigma_def     0.03    0.00 0.01  0.02  0.03  0.03  0.04  0.06    25 1.14
prob_of_draws 0.24    0.00 0.02  0.20  0.23  0.24  0.26  0.29  3297 1.00
```

Output 29: Italian Serie A 2009/2010: model's summary of the dynamic diagonal inflated bivariate Poisson from stan_foot. Posterior estimates for the selected parameters.

In the fits above in Outputs 28 and 29, the prob_of_draws parameter denotes the parameter p to inflate draw occurrences in Equation (5.1). In terms of model comparison with the basic models in Chapter 4, the LOOIC reported by the diagonal-inflated bivariate Poisson model is 2541.8 (with a number of 15.6 estimated parameters), whereas for the dynamic version the LOOIC is 2546.5 (with 18.7 estimated parameters). In any case, these values appear to be much greater than the values reported by the basic models in Section 4.5.7 in Chapter 4: for such reason, diagonal inflation seems to be not very relevant in this dataset. As another empirical confirmation, the bivariate Poisson model is able to adequately capture the draws as shown in Output 27 in the previous Chapter.

5.1.2 Skellam

In some circumstances it could be convenient to work with the goal (score)-difference $Z = Y_1 - Y_2$, i.e. the difference of the goals scored by the two competing teams, in place of the two marginal scores. The new discrete random variable Z is defined on the set of integer numbers $\mathcal{Z} = \{\ldots, -3, -2, -1, 0, 1, 2, 3, \ldots\}$. We suppose that Y_1 and Y_2 jointly follow the bivariate Poisson distribution (4.15), with parameters λ_1, λ_2 and λ_3. Since $P(Z = z) = P(Y_1 - Y_2 = z) = P(X_1 + X_3 - X_2 - X_3 = z) = P(X_1 - X_2 = z)$, the probability function of Z is independent of λ_3 and is the same as that derived from two conditionally independent Poisson variables. Thus, Z follows a discrete probability distribution known as the Skellam (Skellam, 1946), or Poisson-difference, distribution, denoted by $\text{PD}(\lambda_1, \lambda_2)$, whose analytical form is given by:

$$f_Z(z) = \Pr(Z = z) = \exp\{-(\lambda_1 + \lambda_2)\} \left(\frac{\lambda_1}{\lambda_2}\right)^{z/2} I_z\{2\sqrt{\lambda_1 \lambda_2}\}, \qquad (5.2)$$

where $z = \ldots, -3, -2, -1, 0, 1, 2, 3, \ldots$, and $I_r(x)$ denotes the modified Bessel function (Warrick, 1974) defined by:

$$I_r(x) = \left(\frac{x}{2}\right)^r \sum_{k=0}^{\infty} \frac{(x^2/4)^k}{k!\Gamma(r+k+1)}. \tag{5.3}$$

Karlis and Ntzoufras (2003, 2009) propose then to model the goal-difference between the football scores for match i as:

$$\begin{aligned} Z_i|\lambda_{i1}, \lambda_{i2} &\sim \text{PD}(\lambda_{i1}, \lambda_{i2}), \\ \log(\lambda_{i1}) &= \mu + home + att_{h_i} + def_{a_i}, \\ \log(\lambda_{i2}) &= \mu + att_{a_i} + def_{h_i}, \end{aligned} \tag{5.4}$$

where the specification for the scoring rates λ_{i1} and λ_{i2} in terms of offensive and defensive strengths and home effect is the same as for the double Poisson model in (4.2). Identifiability constraints (4.10)–(4.14) can be applied for this model analogously as for the other Poisson models.

Although the Skellam distribution was originally derived as the difference between two conditionally independent Poisson random variables, since the distribution of Z does not depend on the correlation parameter λ_3, treating the number of goals independently for each team leads to an overestimation of model parameters (Karlis and Ntzoufras, 2003). In fact, we should keep in mind that, since the parameters λ_1 and λ_2 are estimated from the marginal distributions, the covariance parameter λ_3 is confounded. For such a reason, we should examine the effects from a model misspecification: if the true underlying model is the bivariate Poisson model, but we use instead the double Poisson model, we assume that the difference follows $Z = Y_1 - Y_2 \sim \text{PD}(\lambda_1 + \lambda_3, \lambda_2 + \lambda_3)$ instead of the correct $Z \sim \text{PD}(\lambda_1, \lambda_2)$. As explained by Karlis and Ntzoufras (2003), the primary effect of this misspecification is that the probability of a draw under a bivariate Poisson model is larger than the corresponding probability under the double Poisson model even if λ_3 is quite small, such as 0.1, which is about the observed covariance in football; moreover, the draw probabilities increase for higher values of λ_3. This fact is an empirical confirmation that the bivariate Poisson model, if compared with the double Poisson, improves over the draw inflation too. To sum up, although the type of the goal difference distribution will be the same regardless of the existence or the type of association between the two variables, this does not imply that the parameter estimates and their interpretation will be the same. As remarked by Karlis and Ntzoufras (2006), the Skellam model can be effectively estimated by use of MCMC methods with sampling augmentation schemes.

5.1.2.1 Implementation in footBayes

In footBayes the user can estimate a Skellam model by using both the maximum likelihood and the Bayesian approach via the following code in Code Snippet 32. The summary is reported in Output 30.

Code Snippet 32 Italian Serie A 2009/2010: Skellam model.

```
### Skellam

## Static
skellam_stan <-  stan_foot(data = "italy_2009",
                           model= "skellam",
                           home_effect = TRUE)
print(skellam_stan, pars = c("home", "sigma_att", "sigma_def"))
```

```
Summary of Stan football model
------------------------------

Posterior summaries for model parameters:
          mean se_mean   sd 2.5%  25%  50%  75% 97.5% n_eff Rhat
home      0.32       0 0.05 0.22 0.29 0.33 0.36  0.43  3886 1.00
sigma_att 0.19       0 0.08 0.05 0.13 0.18 0.23  0.36   415 1.00
sigma_def 0.20       0 0.09 0.05 0.14 0.19 0.26  0.39   356 1.01
```

Output 30: Italian Serie A 2009/2010: model's summary for the Skellam model from stan_foot. Posterior estimates for the selected parameters.

A dynamic Skellam model can be easily fitted as documented in Code Snippet 33. The model's summary is the reported in Output 31.

Code Snippet 33 Italian Serie A 2009/2010: dynamic Skellam model.

```
## Dynamic

skellam_stan_dyn <-  stan_foot(data = italy_2009,
                               model= "skellam",
                               home_effect = TRUE,
                               dynamic_type = "weekly")
print(skellam_stan_dyn, pars = c("home", "sigma_att", "sigma_def" ))
```

The home parameter for the home-effect in Outputs 30 and 31 is estimated to be about 0.32 according to both the modelling version, the static and the dynamic one. Similarly to what happened for the dynamic models in Chapter 4, the standard deviations parameters for the attacking and defensive skill just reached a partial convergence, being the Gelman-Rubin statistic slightly higher than 1.1: in this case, augmenting the number of HMC iterations could help.

```
Summary of Stan football model
------------------------------

Posterior summaries for model parameters:
          mean se_mean   sd 2.5%  25%  50%  75% 97.5% n_eff Rhat
home      0.32    0.00 0.05 0.21 0.29 0.32 0.36  0.43   785 1.01
sigma_att 0.04    0.01 0.01 0.02 0.03 0.03 0.04  0.06     4 1.53
sigma_def 0.05    0.01 0.02 0.02 0.04 0.05 0.06  0.08     6 1.55
```

Output 31: Italian Serie A 2009/2010: model's summary for the dynamic Skellam model from stan_foot. Posterior estimates for the selected parameters.

The attacking and defensive skills under the static Skellam and the dynamic Skellam model are plotted in Figures 5.1 and 5.2, respectively, as obtained by using the foot_abilities function: the results appear to be quite

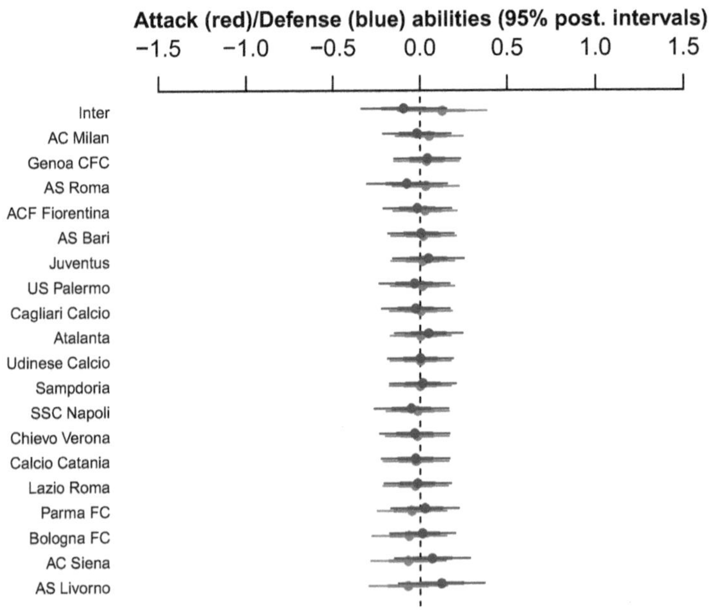

FIGURE 5.1
Italian Serie A 2009/2010: 50% (thicker segments) and 95% (thinner segments) credible intervals for attacking (red lines) and defensive (blue lines) team-specific abilities in the static Skellam model.

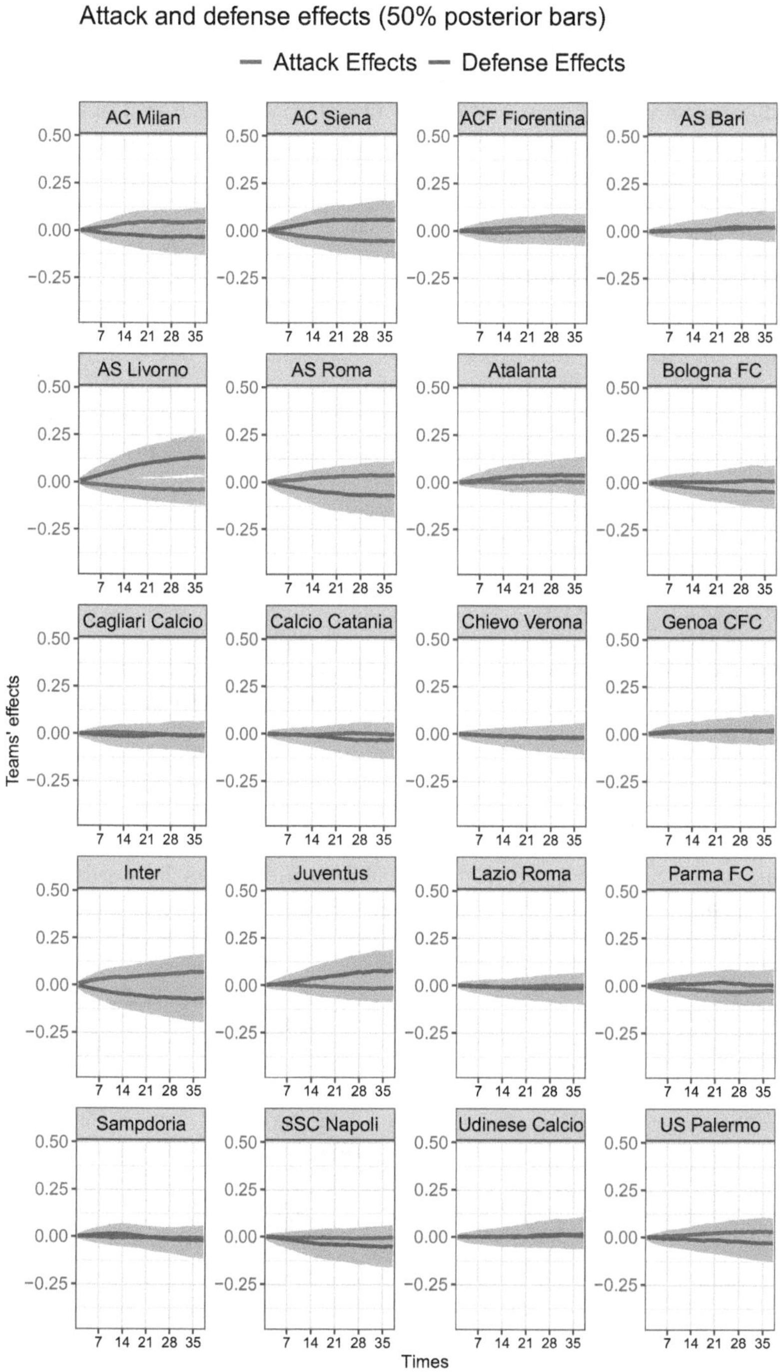

FIGURE 5.2
Italian Serie A 2009/2010: 50% credible intervals (grey ribbons) for attacking (red curves) and defensive (blue curves) team-specific abilities in the dynamic Skellam model.

consistent with those already discussed in Chapter 4 for the basic models, although the amount of shrinkage is greater in the Skellam models: this happens because the Skellam model captures the team-specific abilities in terms of the goal difference and not in terms of number of goals, which makes their estimation somehow less precise—a draw 0-0 counts equally as a draw 4-4 here.

5.1.3 Zero-Inflated Skellam

As suggested by Karlis and Ntzoufras (2009), analogously as for the bivariate Poisson model, the Skellam model (5.4) could be extended to better capture draws' occurrences: the zero-inflated version of the Skellam distribution could be in fact proposed to model an eventual excess of draws in the data. Hence, we can define the zero-inflated Skellam/Poisson difference (ZPD) probability function as:

$$f_Z(z) = \begin{cases} p + (1-p)\text{PD}(z|\lambda_1, \lambda_2), & z = 0 \\ (1-p)\text{PD}(z|\lambda_1, \lambda_2), & z \neq 0, \end{cases} \tag{5.5}$$

where $p \in (0, 1)$ is the draw inflation probability. In statistical notation, we denote with $Z_i \sim \text{ZPD}(\lambda_{i1}, \lambda_{i2}, p)$ the goal difference distributed with a ZPD density function.

5.1.3.1 Implementation in footBayes

The Bayesian estimation for a zero-inflated Skellam model can be produced in footBayes through the following code reported in Code Snippet 34. The model's summary is reported in Output 32.

Code Snippet 34 Italian Serie A 2009/2010: ZPD model.

```
## Zero-inflated Skellam
# Static
zeroinfl_skellam_stan <-  stan_foot(data = italy_2009,
                                     model= "zero_infl_skellam",
                                     home_effect = TRUE)
print(zeroinfl_skellam_stan, pars = c("home", "sigma_att",
                                        "sigma_def", "prob_of_draws"))
```

```
Summary of Stan football model
------------------------------

Posterior summaries for model parameters:
              mean se_mean   sd 2.5%  25%  50%  75% 97.5% n_eff Rhat
home          0.34       0 0.06 0.23 0.30 0.34 0.38  0.44  3727 1.00
sigma_att     0.19       0 0.08 0.05 0.13 0.19 0.24  0.36   416 1.01
sigma_def     0.20       0 0.09 0.05 0.14 0.20 0.26  0.39   435 1.01
prob_of_draws 0.03       0 0.02 0.00 0.01 0.03 0.04  0.08  4051 1.00
```

Output 32: Italian Serie A 2009/2010: model's summary of the zero-inflated Skellam model from stan_foot. Posterior estimates for the selected parameters.

The user can easily fit a Bayesian dynamic zero-inflated Skellam model through the code contained in Code Snippet 35. The model's summary is reported in Output 33.

Code Snippet 35 Italian Serie A 2009/2010: dynamic ZPD model.

```
# dynamic
zeroinfl_skellam_stan_dyn <-  stan_foot(data = italy_2009,
                                    model= "zero_infl_skellam",
                                    home_effect = TRUE,
                                    dynamic_type = "weekly")
print(zeroinfl_skellam_stan_dyn, pars = c("home", "sigma_att",
                                      "sigma_def", "prob_of_draws"))
```

```
Summary of Stan football model
------------------------------

Posterior summaries for model parameters:
              mean se_mean   sd 2.5%  25%  50%  75% 97.5% n_eff Rhat
home          0.34       0 0.06 0.23 0.30 0.34 0.38  0.45   613 1.01
sigma_att     0.02       0 0.01 0.01 0.01 0.02 0.03  0.05     6 2.01
sigma_def     0.05       0 0.02 0.02 0.04 0.05 0.06  0.08    12 1.46
prob_of_draws 0.03       0 0.02 0.00 0.01 0.03 0.04  0.08  3272 1.00
```

Output 33: Italian Serie A 2009/2010: model's summary of the dynamic zero-inflated Skellam model from stan_foot. Posterior estimates for the selected parameters.

As it is evident from the Outputs 32 and 33, the home effect home for the dynamic version is basically the same as for the basic Skellam model, whereas the prob_of_draws parameter appears to be very close to zero, suggesting a low evidence of draw inflation.

5.1.4 Student-t model

As mentioned with regard to the Skellam model in the previous sections, modelling the goal difference in place of the marginal scores could yield some advantages. First, one could implicitly assume scores' correlation without a direct specification; second, modelling only one single process is simpler than modelling two conditionally independent processes or a bivariate structure; third, the goal difference is defined on the support $\mathcal{Z} = \{\ldots, -3, -2, -1, 0, 1, 2, 3, \ldots\}$, for such a reason it could be approximated by a continuous distribution with support in $\mathbb{R}$, such as a Gaussian distribution. However, approximating the score difference with a continuous distribution could oversimplify the problem, since the knowledge obtained from the score difference is less sophisticated than the knowledge arising from the marginal scores: for instance, a draw could be caused by one, two, three, etc. scores for each of the two competing teams. Thus, a model for the goal difference could be less effective in estimating the offensive and defensive strengths, as already mentioned in Section 5.1.2.

By following this streamline, fitting a continuous model for the goal difference could represent a sound alternative method with respect to a discrete distribution. Gelman (2014) and Kharratzadeh (2017) suggest the use of a student-t distribution with $\nu \in \mathbb{R}^+$ degrees of freedom, location $\mu \in \mathbb{R}$, and scale $\sigma \in \mathbb{R}^+$ as follows:

$$f_Z(z) = \frac{\Gamma((\nu+1)/2)}{\Gamma(\nu/2)} \frac{1}{\sqrt{\nu\pi}\sigma} \left(1 + \frac{1}{\nu}\left(\frac{y-\mu}{\sigma}\right)^2\right)^{-(\nu+1)/2}. \tag{5.6}$$

Thus, we could model the score difference Z_i for match i, $i = 1, \ldots, n$ as follows:

$$\begin{aligned} Z_i|\nu, \mu_i, \sigma &\sim t_\nu(\mu_i, \sigma), \\ \mu_i &= \mathit{home} + \mathit{ability}_{h_i} - \mathit{ability}_{a_i}, \end{aligned} \tag{5.7}$$

where h_i and a_i team denote as usual the home and the away team in match i, respectively; $\mathit{ability}_k$ measures the global ability for the k-th team, and *home* is the usual home effect. Kharratzadeh (2017) frames model (5.7) in the Bayesian framework, by proposing the following weakly informative prior distributions (Gelman et al., 2008):

$$\begin{aligned} \mathit{ability}_k &\sim N(\beta, \sigma_a^2), \\ \nu &\sim \text{Gamma}(2, 0.1), \\ \sigma &\sim N(0, 5^2), \\ \sigma_a &\sim N(0, 1), \\ \mathit{home} &\sim N(0, 1), \\ \beta &\sim N(0, 1). \end{aligned} \tag{5.8}$$

Analogously to the attack/defence parameters for the Poisson-based models, a STZ identifiability constraint needs to be imposed to the global abilities, such as:

$$\sum_{k=1}^{K} ability_k = 0. \tag{5.9}$$

As mentioned in Chapter 2 through Equation (2.4), it is easy to state a functional relationship between the global abilities of the student-t model (5.7) and the offensive/defensive parameters in the Poisson-based models (4.2), (4.16) as follows:

$$ability_k = (att_k - def_k), \tag{5.10}$$

thus the global ability for the team k is given by the difference between the offensive and the defensive strengths: the higher $att_k - def_k$, and the higher is the global ability for the team—we stress again that the parameter def represents the defensive weakness, and can be seen as the negative strength. From Equations (5.10) one could build some analogous *a posteriori* measures of global abilities for the Poisson-based models as well, by using the estimates for the parameters att and def previously obtained, as remarked by Equation (2.4).

5.1.4.1 Implementation in footBayes

The student-t model can be estimated in footBayes through the following code in Code Snippet 36. The model's summary is reported in Output 34

Code Snippet 36 Italian Serie A 2009/2010: student-t model.

```
## Student-t

# static
student_stan <- stan_foot(data = italy_2009,
                          model= "student_t")
print(student_stan, pars= c("home", "beta", "sigma_a", "sigma_y"))
```

```
Summary of Stan football model
------------------------------

Posterior summaries for model parameters:
        mean se_mean   sd  2.5%   25%   50%   75% 97.5% n_eff  Rhat
home    0.48   0.00  0.07  0.34  0.43 0.48  0.53  0.62  1459  1.00
beta    0.04   0.06  2.57 -4.93 -1.72 0.056 1.81 5.08  1591  1.00
sigma_a 0.64   0.10  0.69  0.03  0.13 0.38  0.95  2.52    43  1.09
sigma_y 1.19   0.00  0.05  1.09  1.16 1.19  1.23  1.29  1840  1.00
```

Output 34: Italian Serie A 2009/2010: model's summary of the student-t model from stan_foot. Posterior estimates for the selected parameters.

The dynamic student-t model allowing for dynamic team-specific abilities could be estimated through the code in Code Snippet 37. The model's summary is reported in Output 35.

Code Snippet 37 Italian Serie A 2009/2010: dynamic student-t model.

```
# dynamic
student_stan_dyn <- stan_foot(data = italy_2009,
                              model= "student_t",
                              dynamic_type = "weekly")
print(student_stan_dyn, pars= c("home","beta", "sigma_a", "sigma_y"))
```

```
Summary of Stan football model
------------------------------

Posterior summaries for model parameters:
        mean se_mean   sd  2.5%   25%   50%  75% 97.5% n_eff  Rhat
home    0.48  0.00   0.07  0.35  0.43  0.48 0.52  0.60  2729  1.00
beta    0.08  0.04   2.46 -4.84 -1.55  0.04 1.69  4.94  3200  1.00
sigma_a 0.17  0.09   0.16  0.02  0.05  0.09 0.27  0.60     3  2.27
sigma_y 1.18  0.00   0.05  1.08  1.14  1.18 1.21  1.29  1730  1.00
```

Output 35: Italian Serie A 2009/2010: model's summary of the dynamic student-t model from Stan.

The global abilities in Equation (5.8) for the static student-t and the dynamic student-t model are plotted in Figure 5.3 and 5.4, respectively, through the foot_abilities function: by inspecting both the figures, one could realize that higher abilities correspond to the best teams (Inter, AS Roma, AC Milan, Sampdoria), whereas lower abilities correspond to the weakest teams (Livorno, Siena, Atalanta) in the league.

5.2 Model comparison between goal-difference models

We can perform a final model comparison for the goal difference models: in fact, we must take this task separated from the comparison between the goal-based models in Chapter 4 and the diagonal-inflated bivariate Poisson model in Section 5.1.1 because we deal with models based on distinct response variables, the number of scores and the goal difference, respectively. Thus, one could still use the loo package to perform this task as shown in Code Snippet 38. The results of this comparison are shown in Table 5.1.

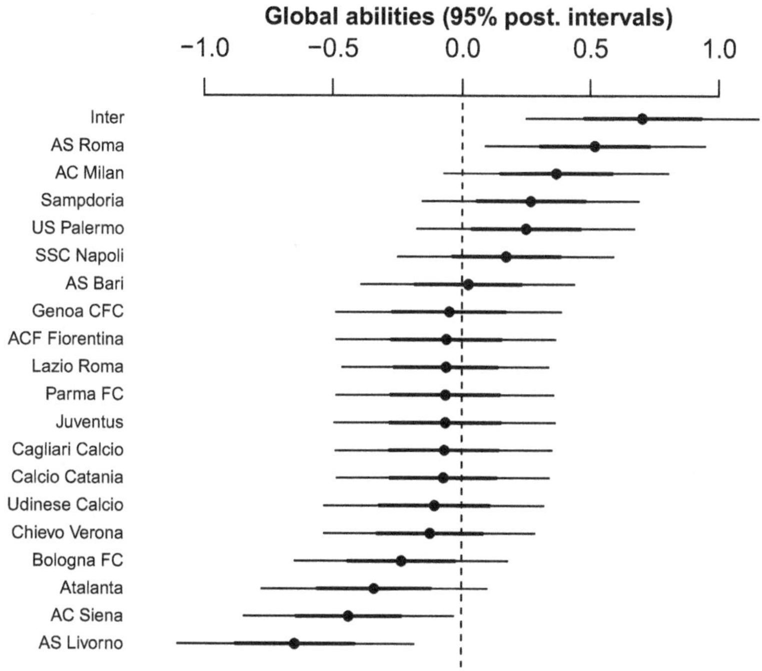

FIGURE 5.3
Italian Serie A 2009/2010: 50% (thicker lines) and 95% (thinner lines) credible intervals for the global abilities in the static student-t model.

TABLE 5.1
Italian Serie A 2009/2010: comparisons via LOOIC for the goal difference models. The second column reports the effective number of parameters as estimated through the procedure

Model	Eff. parameters	LOOIC
Student-t	17.2	1328.6
Dynamic student-t	28.5	1332.7
Skellam	6.5	1380.7
Zero-infl. skellam	7.0	1382.4
Dynamic Skellam	6.9	1383.2
Dynamic zero-infl. Skellam	7.4	1384.8

The static student-t model in Table 5.1 attains the lowest LOOIC value, 1328.6, whereas the dynamic version of the zero-inflated Skellam model yields the highest value, 1384.8: from this comparison as well, we can claim that

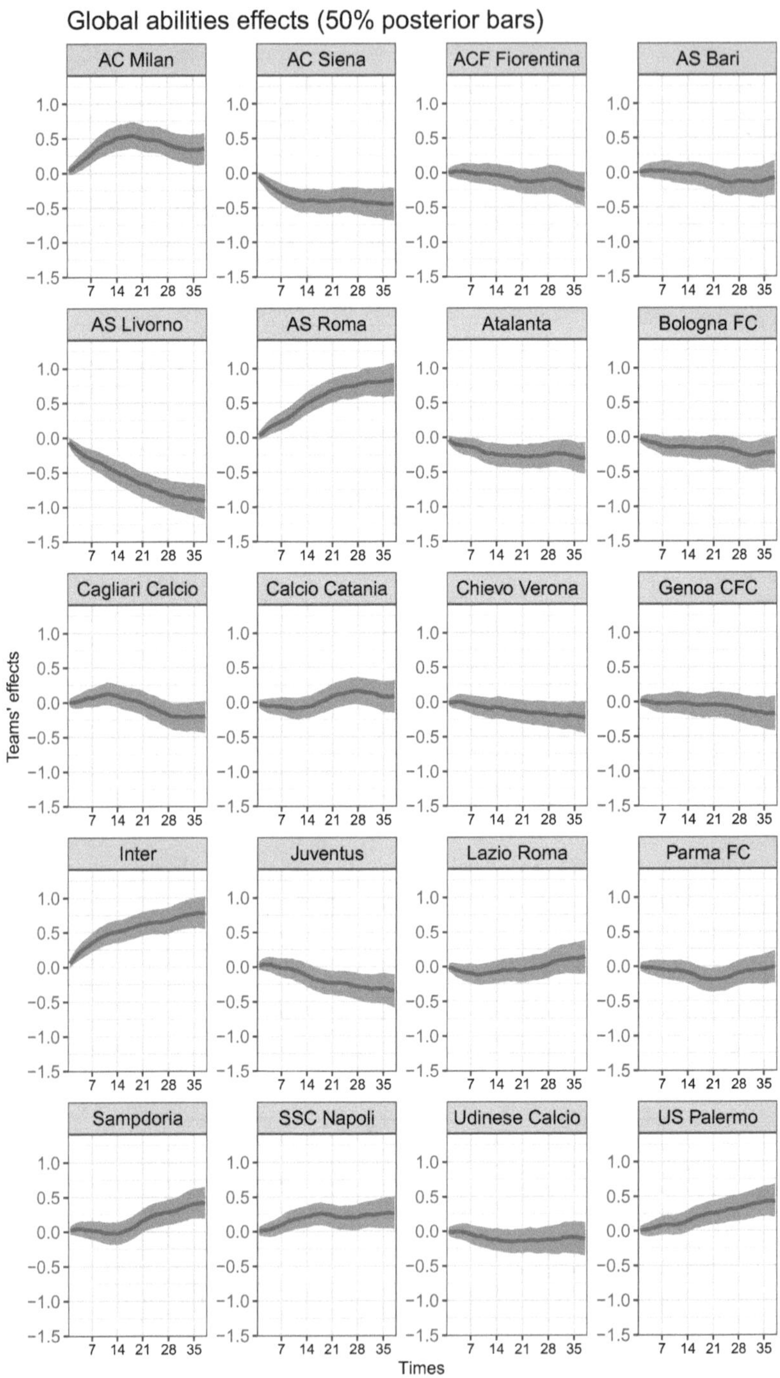

FIGURE 5.4
Italian Serie A 2009/2010: 50% credible intervals for the global abilities in the dynamic student-t model.

Code Snippet 38 Italian Serie A 2009/2010: goal-difference models.

```
### model comparisons
## LOOIC, loo function

# extract pointwise log-likelihood

# static skellam
log_lik_skellam <- extract_log_lik(skellam_stan$fit)
 # dynamic skellam
log_lik_skellam_dyn <- extract_log_lik(skellam_stan_dyn$fit)
# static zeroinfl skellam
log_lik_zeroinfl_skellam <- extract_log_lik(zeroinfl_skellam_stan$fit)
# dynamic zeroinfl skellam
log_lik_zeroinfl_skellam_dyn
                        <- extract_log_lik(zeroinfl_skellam_stan_dyn$fit)
# static student
log_lik_student <- extract_log_lik(student_stan$fit)
# dynamic student
log_lik_student_dyn <- extract_log_lik(student_stan_dyn$fit)

# compute loo

loo_skellam <- loo(log_lik_skellam)
loo_skellam_dyn <- loo(log_lik_skellam_dyn)
loo_zeroinfl_skellam <- loo(log_lik_zeroinfl_skellam)
loo_zeroinfl_skellam_dyn <- loo(log_lik_zeroinfl_skellam_dyn)
loo_student <- loo(log_lik_student)
loo_student_dyn <- loo(log_lik_student_dyn)

# compare six loo

compare(loo_skellam, loo_skellam_dyn,
        loo_zeroinfl_skellam, loo_zeroinfl_skellam_dyn,
        loo_student, loo_student_dyn)
```

according to this Serie A dataset *the simpler the better*: the student-t model with static team-specific abilities performs better than any other more complicated model. In general, the user should compare these model performances on other alternative datasets.

5.3 Adding covariates

In the current literature about football analytics the inclusion and use of covariates has not been always fully addressed; in fact, the most commonly adopted and well-known football models are usually "vanilla" versions—see Chapter 2, Sections 2.1.2 and 2.1.3 for further details— that just use the teams' indicators and their latent abilities without considering any extra explanatory information for modelling and predicting the final football outcomes. However, the use of some relevant covariates regarding the matches, teams, and players along the season could dramatically improve the goodness of fit and offer new insights for the modern football modelling strategies. Models with covariates could in fact benefit from using information from earlier seasons, from experts such as bookmaker companies, or directly use some *tracking* data at the individual level, such as the number of passes, the number of shots, the position, and so on.

In the recent years, propose to include some among the following covariates in the well-known Poisson-based workflow:

- average numbers of goals scored and conceded by every single team, indexed by period prior to the current match, season and division;
- teams' average recent results;
- match importance;
- geographical distance between the home stadium and the away stadium;
- goals scored/conceded in the most recent matches.

Following the arguments in Chapter 2, Section 2.1.1, we denote the covariates as follows:

- $X^{(1)}$: $n \times K$ matrix containing the home team/"team one" covariates for each match-team pair (i, j);
- $X^{(2)}$: $n \times K$ matrix containing the away team/"team two" covariates for each match-team pair (i, j);
- U: other covariates (still in a matrix form), not directly related to the teams.

The inclusion of these covariates in the Poisson-based models is straightforward, since it is sufficient to specify some suited functions f_1, f_2 (for instance, the exponentiated linear predictors) such that $\lambda_{i\cdot} = f_{\cdot}(X^{(1)}, X^{(2)})$. Alternative models may be formulated by varying the functional shape and the set of the covariates in the functions f_1, f_2.

The investigation of candidate covariates likely to improve the fit of the vanilla football models is out of the scopes of this book. Moreover, the advent of the so called *on-field* covariates—running distance, ball possession, number of passes, etc—and the current availability of *tracking* data partially transformed the football analytics field in a big-data framework with a huge number of potential drivers for the final match outcome. For a quick overview, consult the works of Carpita et al. (2015), Groll et al. (2018a), and Groll et al. (2021).

However, we feel there is still a lack of a deep theoretical awareness to connect this huge amount of data with the final match results in order to rely on a coherent statistical modelling workflow. For such a reason, we prefer to focus on team indicators and latent abilities and give a transparent overview of the vanilla model construction, by inviting the interested reader to investigate the potentiality and the inclusion of more and more explanatory variables in the models. As usually remarked—see Chapter 2, Section 2.1.4—a well-known issue with the use of further variables is that the attacking and defensive abilities of the Poisson-based models are usually strongly confounded with the potentially relevant covariates; moreover, prediction with additional covariates may be improved but not as much as one would expect because the abilities capture similar qualities. For tournaments that do not have a round-robin format and where the teams are separated in different groups, the vanilla models will not work until we have a cross over between different groups; in this case, models with covariates perform definitely better. As we will see in Chapter 6, these tournaments require also a covariate describing the overall prior ability of the team, such as the FIFA/UEFA ranking coefficient or any other ranking measure.

Unlike for what happens in national/domestic leagues, such as the Serie A, Premier League, Ligue One, and so on, where the rosters are observed for many match-days week after week, some critical issues arise in international matches, such as those from the Euro Cup, the World Cup or the America's Cup:

- the national rosters may highly vary from match to match, depending on the match type—qualifier, friendly, cup's match—which makes any statistical analysis more unreliable and noisier;
- the number of matches between national teams is much lower than the number of matches registered in domestic leagues;
- compared to domestic leagues, national teams do not play according to a round-robin style. For instance, to gain the qualification for the next World Cup, the best European national teams such as France, Germany or Italy are grouped in different sub-tournaments and expected to play against much lower-level teams, but not to—or rarely—compete one against the other. This aspect makes the statistical analysis about national teams intrinsically *incomplete*;
- national teams do not have any kind of *economic budget*, which instead is highly correlated with teams' performance in domestic leagues;
- international tournaments are usually much more surprising than domestic leagues. Having the best-rated players does not always imply to be the best candidate to win an international tournament;

- national teams are rated according to an official ranking, the FIFA/Coca-Cola world ranking, which monitors national teams performance and lists the best teams by assigning some points to the recent matches results. We will thoroughly use these rankings in Chapter 6 for case studies regarding the Euro and the World Cup.

When the focus is the prediction of international matches, the most natural covariate is represented by the FIFA ranking, eventually rescaled. Specify a bivariate Poisson model such that the FIFA ranking difference for the i-th match $x_i = \text{rank}_{h_i} - \text{rank}_{a_i}$ between the competing teams is added/subtracted to the two scoring rates.

Instead, in a domestic league with an home-effect, introducing a ranking measure—note that FIFA rankings are not available for domestic leagues, and teams such as Inter, AS Roma, and so on, thus the user could supply alternative ranking measures—would lead to extend the vanilla log-linear scores in (2.3):

$$\begin{aligned}\log(\lambda_{1i}) &= \mu + home + att_{h_i} + def_{a_i} + \frac{\gamma}{2}x_i,\\ \log(\lambda_{2i}) &= \mu + att_{a_i} + def_{h_i} - \frac{\gamma}{2}x_i.\end{aligned} \tag{5.11}$$

Alternatively, one could consider the inclusion of a general smoothing function $f(x_i)$ to capture any kind of non-linearity in the ranking difference covariate. Various tests showed that, with regard to international matches data, the coefficient γ is usually highly statistically significant and improves over the fit of the vanilla model not including this covariate. For more details on this topic, we refer the interested reader to the work of Macrì Demartino et al. (2024).

5.3.0.1 Implementation in footBayes

In footBayes the user can include a static or a dynamic ranking measure in the model fit through the optional argument ranking. In case of domestic leagues such as the Serie A, an appropriate measure of ranking could be represented by the rank position each team achieved at the end of the previous season. An example is included in Code Snippet 39, and the corresponding model summary is given in Output 36.

NOTE: the inclusion of ranking measures in the function stan_foot can be performed by supplying through the ranking argument either a data frame containing the teams' ranking measures for distinct match-days or an object of the class "btdFoot" containing the rankings as previously computed by the Bradely-Terry-Davidson model (Bradley and Terry, 1952; Davidson, 1970; Macrì Demartino et al., 2024) through the btd_foot function—type help(stan_foot), help(btd_foot) in the R console and read the thorough vignette accompanying the package for more details.

Code Snippet 39 Italian Serie A 2009/2010: adding ranking covariate.

```
### Adding covariates
## Ranking based on Serie A 2008-2009 rank positions

ranking_2009 <-  as.data.frame(cbind(rep(1,20),
                               unique(italy_2009$home_team),
                               c(17, 14, 1, 15, 5,
                                 2, 10, 20, 8, 7,
                                 18, 3, 6, 11, 9,
                                 16, 4, 12, 19, 13 )))
colnames(ranking_2009) <- c("periods", "team", "rank_points") # rename!
ranking_2009$rank_points <-
      as.numeric(as.vector(ranking_2009$rank_points)) # numeric check

# HMC fit
dp_stan_rank <- stan_foot(data = italy_2009,
                          model="double_pois",
                          home_effect = TRUE,
                          ranking = ranking_2009) # dp + ranking
print(dp_stan_rank, pars = c("home", "gamma", "sigma_att", "sigma_def"))
```

```
Summary of Stan football model
------------------------------

Posterior summaries for model parameters:
           mean se_mean   sd  2.5%   25%   50%   75% 97.5% n_eff Rhat
home       0.41    0.00 0.04  0.33  0.38  0.41  0.44  0.49  4224 1.00
gamma     -0.45    0.00 0.13 -0.71 -0.53 -0.44 -0.36 -0.17   877 1.00
sigma_att  0.09    0.01 0.05  0.01  0.05  0.08  0.12  0.21   100 1.05
sigma_def  0.13    0.00 0.06  0.02  0.09  0.13  0.17  0.25   272 1.01
```

Output 36: Italian Serie A 2009/2010: model's summary of the double Poisson model with rankings from stan_foot. Posterior estimates for the selected parameters.

A comparison in terms of LOOIC with the simple double Poisson model is performed in Code Snippet 40 and reported in Output 37.

Code Snippet 40 Italian Serie A 2009/2010: other model comparisons.

```
# extract pointwise log-likelihood
log_lik_dp_rank <- extract_log_lik(dp_stan_rank$fit) # static dp + rank.
loo_dp_rank <- loo(log_lik_dp_rank)
loo_dp_rank
```

```
Computed from 4000 by 380 log-likelihood matrix.

         Estimate   SE
elpd_loo  -1067.3 16.7
p_loo        14.2  0.8
looic      2134.7 33.3
------
MCSE of elpd_loo is 0.1.
MCSE and ESS estimates assume independent draws (r_eff=1).

All Pareto k estimates are good (k < 0.7).
See help(''pareto-k-diagnostic'') for details.
```

Output 37: Italian Serie A 2009/2010: LOOIC for the double Poisson model with the rankings.

The parameter γ associated to the ranking difference in Output 36 is estimated to be about -0.45 and its 95% credible interval does not contain the zero: according to the Equation (5.11), as a matter of illustration the ranking difference between the home team Inter—the Serie A 2008/2009 winner—and the away team Bari—the first team promoted from the second Italian league, the Serie B 2008/2009—is $1 - 18 = -17$. The ranking is automatically centred and scaled by the stan_foot function to have mean 0 and standard deviation 0.5. Thus, since $x_i = \text{rank}_{h_i} - \text{rank}_{a_i} = -0.803 + 0.634 = -0.169$ in this imaginary match, then $x_i \times \hat{\gamma}/2 = -0.169 \times -0.225 = 0.038$, which means that the multiplicative effect on the average scores implied by this ranking difference for Inter is about $\exp\{0.038\} \approx 1.039$, whereas is about $\exp\{-0.038\} \approx 0.963$ for Bari. The negative sign of the γ coefficient makes then perfectly sense, being $\exp\{\hat{\gamma}/2\} = \exp\{-0.225\} = 0.799$ the multiplicative effect on the average home-team scores for an unitary difference of 1 in the (scaled) rankings, which also means that for every unitary difference in ranking positions there is a 20.1% decrease of average scores for the home team, and a 25.2% increase of average scores for the away team—note that the higher (lower) x_i, the worse (better) is the home team and the better (worse) is the away team.

The LOOIC in Output 37 is 2134.7, lower than the LOOIC for the double Poisson model reported in Section 4.5.7, 2138.7. Although not fully reported here, the inclusion of such ranking improves the fit in each of the models considered so far: as it is intuitive, the inclusion of a ranking measure covariate overall improves the fit of a football model and is then strongly suggested.

5.4 Additional models

5.4.1 Scaled double Poisson from Dixon and Coles (1997)

Although the bivariate Poisson distribution (4.15) is one of the most natural ways to deal with scores' dependence, Dixon and Coles (1997) argue that the

BP model is not able to represent departures from independence for low scoring games, such as 0-0, 0-1, 1-0, or 1-1. For such a reason they proposed a joint distribution for the pair (Y_1, Y_2) by scaling the double Poisson distribution in (4.1) as follows:

$$f_{Y_1,Y_2}(y_1, y_2) = \Pr(Y_1 = y_1, Y_2 = y_2) = \tau_{\lambda_1,\lambda_2}(y_1, y_2)\frac{\lambda_1^{y_1}\exp\{-\lambda_1\}}{y_1!}\frac{\lambda_2^{y_2}\exp\{-\lambda_2\}}{y_2!}, \tag{5.12}$$

where λ_1 and λ_2 represent the marginal means, whereas the scaling parameter τ is defined as follows:

$$\tau_{\lambda_1,\lambda_2}(y_1, y_2) = \begin{cases} 1 - \lambda_1\lambda_2\rho, & \text{if } y_1 = y_2 = 0, \\ 1 + \lambda_1\rho & \text{if } y_1 = 0, y_2 = 1, \\ 1 + \lambda_2\rho & \text{if } y_1 = 1, y_2 = 0, \\ 1 - \rho & \text{if } y_1 = y_2 = 1, \\ 1 & \text{otherwise.} \end{cases} \tag{5.13}$$

The parameter ρ satisfies

$$\max(-1/\lambda_1, -1/\lambda_2) \leq \rho \leq \min(1/\lambda_1\lambda_2, 1),$$

and enters as a dependence parameter measuring the correlation between the scores, such that $\rho = 0$ corresponds to scores' independence, but otherwise the independence assumption is perturbed for events with $y_1 \leq 1$ and $y_2 \leq 1$. The authors proved that the corresponding marginal distributions remain Poisson with means λ_1 and λ_2, respectively. To achieve identifiability, the constraint (4.14) is applied. By assuming the parametrization (4.3), the likelihood function takes the following form (proportionality constants have been dropped to ease the readability):

$$\begin{aligned} \mathcal{L}(\alpha, \beta, \gamma, \rho; y_1, y_2) &= \prod_{i=1}^{n} \tau_{\lambda_{1i},\lambda_{2i}}(y_{i1}, y_{i2})\lambda_{i1}^{y_{i1}}\exp\{-\lambda_{i1}\}\lambda_{i2}^{y_{i2}}\exp\{-\lambda_{i2}\} \\ &= \prod_{i=1}^{n} \tau_{\lambda_{i1},\lambda_{i2}}(y_{i1}, y_{i2})(\gamma\alpha_{h_i}\beta_{a_i})^{y_{i1}}\exp\{-(\gamma\alpha_{h_i}\beta_{a_i})\}\times \\ &\quad (\alpha_{a_i}\beta_{h_i})^{y_{i2}}\exp\{-(\alpha_{a_i}\beta_{h_i})\}, \end{aligned} \tag{5.14}$$

where λ_1 and λ_2 are defined as in Equation (4.2)—or, equivalently, as in (4.3)—except for the fact that $\mu = 0$, then $\delta = \exp\{\mu\} = 1$. Inferential conclusions are provided in Dixon and Coles (1997) by numerical optimization of the likelihood function (5.14).

NOTE: the implementation of the scaled double Poisson is not guaranteed by the footBayes package, but one could fit this model by using the regista package, as shown in the next section.

5.4.1.1 Implementation in regista

The user could fit a scaled double Poisson model through the dixoncoles function in the regista package—downloaded from GitHub. The maximum likelihood estimates for the attacking and the defensive abilities are shown in Figure 5.5. In comparison with the static team-specific abilities from the bivariate Poisson model in Output 13 and the Skellam model in Figure 5.1, we could note a general agreement: stronger teams, such as Inter, AS Roma and AC Milan are associated with the highest abilities, conversely weaker teams, such as AS Livorno, Chievo Verona, and Atalanta exhibit the worst strengths.

Code Snippet 41 Italian Serie A 2009/2010: scaled double Poisson model.

```
## Scalde double Poisson
install_github("torvaney/regista")

library(regista)
library(arm)

dc_fit <- dixoncoles(hgoal = italy_2009$home_goals,
                     agoal = italy_2009$away_goals,
                     hteam = factor(italy_2009$home_team),
                     ateam = factor(italy_2009$away_team),
                     data = italy_2009) # dixon-coles model
```

5.4.2 The count Weibull model

So far, the majority of the models started from an assumption that the underlying process was a time-homogeneous Poisson process. In such a process, time between events (goals) follows an exponential distribution, due to the well-known relationship between the Poisson and the exponential distribution. Exponential inter arrival times actually imply that the hazard of a goal remains the same in the entire match, which can be counter-intuitive. For example, one would expect that the team that it is behind in score would try harder to equalize and this may change the inter-arrival distribution. This assumption, however, is a convenient assumption due to lack of alternatives

In Boshnakov et al. (2017) the inter-arrival times are assumed to follow an independent and identically distributed Weibull distribution. They based on the results from McShane et al. (2008).

Let $Y(n)$ be the time at which the n-th event (goal in our case) occurs. Let $E(t)$ denote the number of events that have occurred up until time t. The relationship between inter-arrival times and the number of events is

$$Y(n) \leq t \Leftrightarrow E(t) \geq n$$

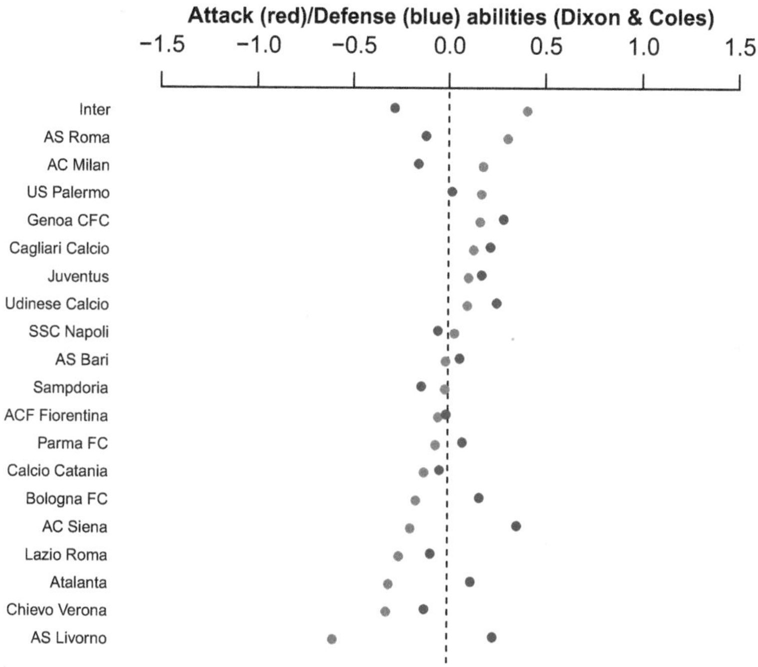

FIGURE 5.5
Italian Serie A 2009/2010: point estimates for the attacking (red points) and defensive (blue points) team-specific abilities in the scaled double Poisson model from Dixon and Coles (1997).

and hence one can derive the discrete probability for the number of events from that of the distribution assumed for the inter-arrival times, namely

$$\begin{aligned}\Pr(E(t) = n) &= \Pr(E(t) \geq n) = \\ &\Pr(E(t) \geq n+1) = \Pr(Y_n \leq t) - \Pr(Y_{n+1} \leq t)\end{aligned} \tag{5.15}$$

In order to find the distribution of $Y(n)$ it suffices to see that $Y(n) = \sum_{i=1}^{n} Y_i$ where Y_i follows the distribution assumed for the inter-arrival times and hence this is the convolution of independent variates from this distribution.

Assuming a Weibull inter-arrival time distribution, the number of goals has probability mass function

$$\Pr(E(t) = x) = \sum_{j=x}^{\infty} \frac{(-1)^{(x+j)}(\lambda t^c)^j \alpha_j^x}{\Gamma(cj+1)}, \tag{5.16}$$

where $\alpha_j^0 = \frac{\Gamma(cj+1)}{\Gamma(j+1)}$, $j = 0, 1, \ldots$, and $\alpha_j^{x+1} = \sum_{m=x}^{j-1} \alpha_m^x \frac{\Gamma(cj-cm+1)}{\Gamma(j-m+1)}$ for $x =$

$0, 1, 2, \ldots$ and $j = x+1, x+2, \ldots$. In (5.16), λ is a rate parameter and c is the shape parameter of the distribution, where the observation unit is the match, which we take as having a duration of one time unit. Thus, λ is the scoring rate per match.

The associated law with the count process has hazard function $h(t) = \lambda c t^{c-1}$ and it varies over time while it can have different shapes for different segments of the match. It can be monotonically increasing for $c > 1$, monotonically decreasing for $c < 1$, or constant (and equal to λ) for $c = 1$. Note that we recover the (time-homogeneous) Poisson process when $c = 1$ and hence the count Weibull distribution is the Poisson distribution.

This brings to a new discrete distribution named *the Weibull count distribution*; it is different from the *discrete Weibull distribution*, which is simply based on the discretization of the Weibull distribution and not in some underlying process like the one described above. If one prefers some other choices of the underlying inter-arrival distribution can see Nadarajah and Chan (2018).

It is also interesting to note that this model handles both over-dispersed data—the mean is smaller than the variance; $c < 1$ and under-dispersed data—the mean is larger than the variance; $c > 1$—naturally, whilst the Poisson count distribution ($c = 1$) can only accommodate equi-dispersed data—the mean is equal to the variance).

Boshnakov et al. (2017) use this as the marginal distribution to derive a bivariate model similar to those described in Chapters 4 and 5: a bivariate model can be described using a copula to couple the two variables (number of goals). They use a a Frank copula by including covariates in the rate parameters as usual: the computational steps required to estimate this model can be performed with small effort by using the R library `countr`.

5.4.3 The Copula model

A brief discussion about Copulas

As previously remarked, the existence of some sort of dependence between the goals scored by two teams in a football match is widely accepted. It is also apparent that one needs to account for that in the modelling phase, as described in Chapter 4. However, the exact specification of this dependence is less clear. It can be positive and negative and in general we need to be flexible on that considering models that allow for a wide range of correlation. Even if it is small, it can have an effect in our predictions so perhaps we do not have to ignore it.

As we have remarked in the previous sections, existing models try to fit the correlation either implicitly via the covariates that share common information or explicitly by allowing for a correlation parameter. For example the bivariate Poisson in Maher (1982) and introduced in Section 4.4.2 allows only for positive correlation which has a linear form. Dixon and Coles (1997) show the observed joint frequencies and compare them under the assumption of independence, by claiming that that certain cells deviate and perhaps they introduced the extra terms to account for that. Also the model in Karlis and Ntzoufras (2003) starting from a bivariate Poisson adds extra correlation from the inflation terms in the diagonal.

An alternative way to incorporate dependence is by using copulas. Copulas can produce flexible bivariate (multivariate) distributions with flexible marginal distributions and flexible dependence structure—e.g. we can easily create a bivariate discrete distribution with negative correlation. Since we can have a great variety of different copulas, different dependence structures are possible. Copulas also offer a way of studying scale-free measures of dependence. The cost of using copulas refers to the added complexity which can be relaxed since there are now many available packages.

Copulas are bivariate (multivariate) distributions with uniform marginals. They are fashionable since one can separate the marginal properties from the dependence properties—caution: this is not true for discrete data however—and hence define multivariate models with given marginal properties. They have found increasing application to many disciplines, like biostatistics, finance, hydrology, and literature is increasing fast.

We refer to copulas as distribution functions whose one dimensional margins are uniform.

Definition (Nelsen, 2006): *Let* $\mathbf{I} = (0, 1)$. *A bivariate copula is a function* C *from* $\mathbf{I}^2$ *to* $\mathbf{I}$ *with the following properties:*

1. *For every* u, v *in* $\mathbf{I}$

$$C(u, 0) = 0 = C(0, v) \qquad \textit{and} \qquad C(u, 1) = u, C(1, v) = v$$

2. *For every* u_1, u_2, v_1, v_2 *in* $\mathbf{I}$ *such that* $u_1 \leq u_2$ *and* $v_1 \leq v_2$,

$$C(u_2, v_2) - C(u_2, v_1) - C(u_1, v_2) + C(u_1, v_1) \geq 0.$$

If $C(\cdot, \cdot)$ is considered to be a distribution function of two random variables U and V, the first condition ensures that U and V have uniform marginal distributions. The second condition, often referred to as the rectangular inequality, simply requires that C is a valid distribution function, i.e. $\Pr(u_1 \leq U \leq u_2, v_1 \leq V \leq v_2) \geq 0$.

Note that if the marginal distribution functions are continuous then the copula is unique. In the discrete case, the copula is not unique in general but it still permits the construction of valid parametric statistical models. The difference with the continuous case is that the copula parameter alone does not characterize the dependence between the random variables at play (Genest and Nešlehová, 2007). Nevertheless, since any well-defined copula-based model is a particular instance of a statistical model, the tools and the methods of the latter can be applied to the former. When the marginal distributions are discrete, understanding the dependence structure implied by the fitted copula is somewhat more complex than in the continuous case due to the possibility of ties (equal marginal values). A thorough discussion on the use of copulas for discrete data can be found in Nikoloulopoulos (2013).

For the bivariate case, and using a copula $C(\cdot,\cdot)$ with marginal distributions $F(y_1)$ and $G(y_2)$, the bivariate joint probability mass function $h(y_1, y_2)$ is given by

$$\begin{aligned} f(y_1, y_2) &= C(F(y_1), G(y_2)) - C(F(y_1 - 1), G(y_2)) \\ &- C(F(y_1), G(y_2 - 1)) + C(F(y_1 - 1), G(y_2 - 1)) \end{aligned}$$

The choice of a copula family can be guided by the (dependence) properties of that family. For example, one may seek a dependence structure that is comprehensive, meaning the copula family can model the full range of dependence structure with correlation ranging from –1 to 1. Or one may expect to have a larger correlation for large values, which is known as tail dependence. Different copulas are offering different structures and hence a natural question is which one is better to use.

The model

McHale and Scarf (2011a) propose a copula model for the number of goals scored by opposing teams in international soccer matches. International soccer—i.e. matches between national teams—is not organized into hierarchical leagues and as a result there are games between teams with a much wider variation in ability. Recall that, for example, in the group-stage of the preliminaries of the World Cup, different groups of abilities are formed and one teams from each group is selected. So, weaker teams play against stronger teams and matches between teams of very different abilities are frequently observed.

The bivariate discrete distributions employed are defined through copulas. This allows dependence in the bivariate distribution to be modelled in a flexible manner by specifying a suitable family of copula functions. A first observation was based on the fact that typically matches with larger rank difference between teams have larger correlation and hence there is need for a model that can capture this. Namely, games between closely ranked teams, the overall dependence is low, and that the dependence becomes increasingly negative as the competitiveness of a match decreases.

They used data from 6101 international soccer results for the period 1993-2004. Marginal means are modelled with match related covariates. They considered both Poisson and negative binomial marginal distributions. The negative binomial, with the same overdispersion for each team, is selected based on AIC. Then the two marginals are coupled with a copula. The family of Archimedean copulas are used and finally a Frank copula is chosen. The Frank Copula has a cdf defined by

$$C(u, v) = -\frac{1}{\theta} \log \left\{ 1 + \frac{(e^{-\theta u_1} - 1)(e^{-\theta u_2} - 1)}{e^{-\theta} - 1} \right\},$$

where θ is the copula parameter taking values in $\mathbb{R}-\{0\}$. As θ approaches 0 we get the product copula, which implies independence between the two variables. The Frank copula allows for both negative and positive dependence. Then the copula parameter θ is modelled through a linear function, namely

$$\theta_i = \beta_0 + \beta_1 x_i,$$

where x_i is the difference in the ranking of the two teams as in Equation (5.11), and β_0 and β_1 parameters need to be estimated. A value of $\beta_1 = 0$ implies constant dependence. The marginal means are modelled using past records of the two teams and in particular the number of goals scored by the team and the number of goals conceded in the last eight matches. The number of matches is selected after some selection procedure. Also a home advantage is assumed, together with a dummy for matches in neutral field—as it is typical for large tournaments like the World Cup—plus the ranking difference of the teams. The final results confirm that the dependence is different depending on the competitiveness of the match as indicated by the rank difference.

There are some more papers that make use of copulas in football modelling. As far as modelling the number of goals one can see the works in Lee (1999), which refers to Australian football but this is the first attempt to model scores with copulas. Also the work in Boshnakov et al. (2017) and Barbiero (2020) create bivariate models based on copulas. Another example is given by McHale and Scarf (2007), where the number of shots made from the two teams are considered. The shots typically have much larger correlation which is also negative and hence copulas-based models are of large importance. Finally, Dawson et al. (2007) use copulas for the joint distribution of red and yellow cards.

5.5 Summary and closing remarks of Chapter 5

Modelling the result of a football match can be addressed in many alternative ways: the current chapter provides an overview about some models usually adopted for football modelling which extend the basic model formulations in Chapter 4. Some of these specifications involve the use of goal difference as the new response variable.

The Poisson-based models in Chapter 4 could usually suffer from an underestimation of the number of draws: an inflated bivariate Poisson version is proposed in Section 5.1.1 to inflate the draws' occurrence and provide better estimates. Skellam, zero-inflated Skellam, and student-t models for the goal difference are proposed in Sections 5.1.2, 5.1.3, and 5.1.4, respectively, along with a minimal code to fit them through the footBayes package. A thorough model comparison in terms of leave-one-out predictive information criterion (LOOIC) is proposed in Section 5.2.

Section 5.3 explains how to add covariates, specifically some ranking measures, in the previous models for both the scores and the goal difference models.

The scaled double Poisson model from Dixon and Coles (1997) is introduced in Section 5.4.1.

Going beyond the Poisson-based models and the other models for goal difference, a count Weibull model for the inter-arrival times of the goals is provided in Section 5.4.2. Finally, Section 5.4.3 focuses on the use of copula models to capture scores' dependence from a larger perspective.

6

Modelling international matches: The Euro and World Cups experience

6.1 Data and modelling a knock-out tournament

In Chapters 4 and 5 we showcased the estimation of basic and advanced goal-based football models via the footBayes package with regard to domestic leagues, such as the Serie A 2009/2010. However, as remarked in the previous chapters, modelling a *knock-out* tournament is rather different than modelling a national seasonal competition. Thus, we need to take these considerations in mind when building and fitting a proper statistical model for a Euro, Asia, Africa, or World cup competition. Many factors contribute to create such a distinction, we list here the main ones.

- *Unbalanced number of matches*: in international matches data some teams play more matches, some other teams less. As a consequence, team-specific abilities will be estimated more or less precisely depending on the number of matches.
- *Structure*: the structure of Euro and World cups is rather different than a seasonal championship: a group-stage consisting of four teams is followed by a pure knock-out scheme, usually from the round of sixteen until the final.
- *Training set data*: a relevant question regards the kind, amount, and quality of the training set used to train our models in international matches. Should we consider all the qualifiers and friendly matches played one, two, three, or four years before the Cup? Should instead we consider only the qualifiers and official matches? And, quite importantly, should we consider also the results from the previous Euro/World Cups, eventually played four, eight or even twelve, if not sixteen, years before? Even by assuming that a somehow optimal choice of the training set is made, how to account then for the matches in terms of their relative importance and/or their chronological order? Maybe, one could weight the distinct data sources, for instance by supposing that a friendly match is less important than a qualifier match, and down-weight the matches further back in time, as proposed by Dixon and Coles (1997) and reported in equation (5.14).

DOI: 10.1201/9781003186496-6

- *Individual player evaluation*: national teams do not behave as the football *clubs* that compete in the domestic leagues. In fact, they cannot buy or sell some players during the transfer-market. Then, national teams can exhibit much more rosters' variability than Premier League, Serie A, or Bundesliga teams across time. As an example, consider that in the months that precede an important tournament such as the World Cup or the Euro Cup the national coaches try different players, different schemes or tactical situations during friendly matches or, say, Nation's League matches; however, with high chances many of these attempts are not successful and will not be proposed during the targeted tournament. From a statistical perspective, it is then not trivial to analyze historical data from teams exhibiting large variations in their rosters.

- *Covariate availability*: many national team-level covariates could be relevant in modelling and predicting football outcomes in Euro and World Cups. Groll and Abedieh (2013); Groll et al. (2016, 2018a, 2021) use for instance some economic factors such as the GDP per capita, the population, the odds, some sportive factors, such as the average market value, the FIFA ranking, the UEFA points, and some factors describing the teams' structure, as the number of players abroad, the number of Champion's League players, the difference from optimal age, the age and the nationality of the coach. However, there is no home-effect, since in the international tournaments there is just usually one, or two, hosting team.

- *Extra time*: conversely to what happens in domestic leagues, international football matches in the knock-out stage could last more than 90 minutes. This happens when the regular time finishes with a draw, and the two teams are then asked to play further 30 minutes and, if the draw equilibrium persists, deal with some penalties kicks to elect the match winner. This occurrence makes the modelling procedure more appealing: eventually, a proper model for the extra time could be designed.

6.2 Euro Cup 2020 and World Cup 2022

In the following sections we consider and present some "snapshots" directly obtained from our experience of "modellers" for the Euro Cup, 2020 (but played 2021 for Covid-19 reasons) and the World Cup 2022 hosted by Qatar. The reason to consider the two tournaments together reflects their similarity: after a group-stage phase, consisting of small groups, each constituted by four teams playing one against the other only one time according to a partial round-robin format, there are then the round of sixteen, the quarter of finals, the semifinals and the final.

Although we provide a unique treatment, we want to make some disclaimers. The data collection for these two tournaments could be performed in different ways. In fact, consider that the Euro 2020 teams are more homogeneous and similar if compared with the teams participating in the World Cup 2022, also from a geographic perspective. Keep also in mind that the European teams, in order to achieve the World Cup qualification, do not compete against non-European national teams: this aspect makes the World Cup much noisier and challenging to predict, since the eventual match Germany-Argentina has no occurrences in the past—beyond any eventual friendly match between the two teams.

We present here some results and, more in general, the challenges emerged by modelling on-line the two competitions through a diagonal-inflated bivariate Poisson model—see Equation (5.1) in Chapter 5—designed to better capture the draw occurrences. We want to stress the fact that the whole proposed analysis arises from some subjective choices and may, of course, be changed/updated according to distinct modeller beliefs—actually, we argue that this analysis could represent a nice motivation to fit your own models and eventually improve the predictive results. We remind the interested reader to look at the comprehensive documentation of our modelling experience during the Euro Cup 2020 and the World Cup 2022[1].

6.2.1 Data

When preparing for modelling a knock-out tournament involving many national teams, the choice of the training set is of primary relevance and could dramatically affect the final predictions. Many and crucial questions arise: what kind of matches should we consider? Should we weight the matches in terms of their importance—for instance, a friendly match is less informative and valuable than a qualifier match? And then, how many past years should we consider?

We try to classify the major problems surrounding data collection for a generic knock-out tournament.

- *Temporal*: if we consider the previous national teams' performances as a whole, we incur in the risk of not capturing temporal trends. For instance, consider the Italian national team: they missed the Russia World Cup qualification on November 2017, then they had a very long consecutive pattern of wins and draws in the years 2019-2021, and in 2021 they won the Euro Cup against the English national team. However, they missed again the Qatar World Cup 2022 qualification on March 2022. Thus, it is pretty impossible to wisely use their data from 2017 to 2022 without considering temporal trends.

[1] `https://statmodeling.stat.columbia.edu/2022/11/19/football-world-cup-2022-predictions-with-stan/`.

- *Selection bias*: one could arbitrarily choose to discard some games, such as the friendly matches, since they are usually associated with less relevant information in terms of sportive factors. However, the problem is not avoided: are we sure to not discard some relevant, even if marginal, information? For instance, there are some friendly matches between very old traditional national teams—such as England, France, Germany, Italy, among the others—that are usually played with high intensity and are perceived to be particularly hot for both players and fans. Alternatively, one could decide to weight the matches in terms of their importance, by establishing a sort of ranking: tournaments, qualifiers, and friendly games. In such a way, some matches would be down-weighted when fitting the models. However, it is not straightforward how to fix/estimate the weights, and the procedure could yield a sort of selection bias.

- *Game importance*: it is trivial to claim that a qualifier match could carry much more information than a friendly match, where usually the coach makes some rudimental/preliminary attempts, and the players participate with less intensity. Anyway, it is not trivial to establish the game importance in a clear way, possibly because some qualifiers matches could be much more unbalanced than a friendly match between to top teams.

- *Roster dynamics*: national teams suffer from a long and physiological turnover, since there are no job contracts linking national players with their national teams. The players are usually convoked by the national coaches on the ground of their individual performance in the respective domestic leagues. For this reason, a national roster in 2018 could be dramatically different from a roster of the same national team in 2020.

For these reasons, in modelling both the Euro Cup 2020 and the World Cup 2022 we decided to adopt the following conservative strategy:

- Consider all the national teams matches in the previous four years.

- No game importance accounted in data collection, every match is identically contributing to the final results.

- No temporal account in the matches. Rather, we let the model, as explained in the next section, to account for dynamic trends in the team-specific abilities, as proposed in Chapter 4, Section 4.4.3.

- No account for roster dynamics.

Here below we provide in Code Snippet 42 the data acquisition in R for both the Euro Cup 2020 and the World Cup 2022, along with the FIFA rankings available before the competitions started [2]. A sketch of the data is displayed in Output 38.

[2] `https://inside.fifa.com/fifa-world-ranking`.

Code Snippet 42 Euro Cup 2020 and World Cup 2022: data acquisition.

```
library(footBayes)
library(devtools)
library(dplyr)
library(bayesplot)
library(ggplot2)
library(loo)

### EURO CUP 2020
euro_data <- read.table("euro.csv", sep=",", header = TRUE)
euro_data <- euro_data[,-1]
colnames(euro_data) <- c("periods", "home_team", "away_team",
                         "home_goals", "away_goals")
rankings <- read.csv("ranking_euro.csv", sep=";") # fifa rankings
head(euro_data)

### WORLD CUP 2022
wc_data_train <- read.csv("world.csv", sep=",")
wc_data_train <- wc_data_train[,-1]
colnames(wc_data_train) <- c("periods", "home_team", "away_team",
                             "home_goals", "away_goals",
                             "tournament")
rankings <- read.csv("ranking_world.csv", sep=",") # fifa rankings
rankings <- rankings[,-1]
head(wc_data_train[,-6])
```

```
# Euro Cup
  periods        home_team  away_team home_goals away_goals
1       1       Kazakhstan   Scotland          3          0
2       1 Northern Ireland    Estonia          2          0
3       1      Netherlands    Belarus          4          0
4       1         Slovakia    Hungary          2          0
5       1          Croatia Azerbaijan          2          1
6       1           Israel   Slovenia          1          1

# World Cup
  periods  home_team            away_team home_goals away_goals
1       1     Belize             Barbados          1          0
2       1  Palestine                 Iraq          0          3
3       1    Andorra United Arab Emirates          0          0
4       1   Barbados              Jamaica          2          2
5       1 Bangladesh            Sri Lanka          0          1
6       1      Macau      Solomon Islands          1          4
```

Output 38: Euro 2020 and World Cup 2022 datasets' structure.

6.2.2 Tournaments scheme

The Qatar World Cup 2022 consists of:

- *32 national teams*, 64 matches.
- *8 Group-stages.* Group A: Netherlands, Senegal, Ecuador, Qatar. Group B: England, United States, Iran, Wales. Group C: Argentina, Poland, Mexico, Saudi Arabia. Group D: France, Australia, Tunisia, Denmark. Group E: Japan, Spain, Germany, Costa Rica. Group F: Morocco, Belgium, Croatia, Canada. Group G: Brazil, Switzerland, Cameroon, Serbia. Group H: Portugal, South Korea, Uruguay, Ghana.
- *Knock-out stage*: round of sixteen, quarter of finals, semifinals, finals.

The Euro Cup 2020 consists of:

- *24 national teams*, 51 matches.
- *6 group-stages.* Group A: Italy, Turkey, Wales, Switzerland. Group B: Denmark, Finland, Belgium, Russia. Group C: Netherlands, Ukraine, Austria, North Macedonia. Group D: England, Croatia, Scotland, Czech republic. Group E: Spain, Sweden, Poland, Slovakia. Group F: Hungary, Portugal, France, Germany.
- *Knock-out stage*: round of sixteen, quarter of finals, semifinals, final.

In both the tournaments, the group-stage consists of a partial mini round-robin format, where each team plays against all the others only once—of course, there is no account for the home effect here. The group-stage matches terminate within the regular 90 minutes (plus additional minutes decided by the referee) with a win of the first team, a draw, or a win of the second team. Regarding the World Cup, the first two teams for each group rank are qualified for the knock-out phase—for a total of 16 teams—whereas for the Euro Cup the first two teams in their final group's rank and the best four among those who concluded at the third position are qualified for the knock-out phase—for a total of 16 teams as well. From the knock-out phase, a match ends within the regular time only if one of the two teams is ahead in terms of scores/goals: in case of draw, the match is extended by extra thirty minutes. If the draw equilibrium persists after these thirty minutes, penalties kicks are used to select the winner.

The Euro Cup 2020 has been won by Italy, who defeated England in the final match at the penalties kicks, while in the regular 90 minutes the match ended 1-1. Argentina is instead the World Cup 2022 winner: they beated France at the penalties kicks, the match after the regular time terminated 3-3.

Keep in mind that our models in the following sections were designed to model and predict the match results within the regular 90 minutes (plus extra-times), perhaps we did not properly model the extra time in case of draws and the penalties. However, we feel that the extra time after the regular time could be further modelled: this aspect is out of the scopes of the book and could represent a point for future research.

6.2.3 The rankings

Coca-Cola FIFA rankings[3] represent one of the most well-known proxies describing national teams' strengths. The actual FIFA algorithm is called SUM and works by adding or subtracting some points that are partially determined by the relative strength of the two opponents. The formula of the SUM algorithm for determining the number of FIFA points is specified as follows:

$$\text{Points} = \text{Points}_{before} + I \times (W - W_e), \tag{6.1}$$

where Points_{before} denotes the number of points before the match, I is the match importance defined in a scale from 5 to 60 depending on the match competition, W is the final result of the match (1 point for win, 0.5 for draw and 0 for defeat), and W_e is the expected result of the match, computed as $W_e = 1/(10^{-\text{dr}/600}+1)$; $\text{dr} = \text{Points}_{before,\ team\ A} - \text{Points}_{before,\ team\ B}$. There are then some adjustments of (6.1) for matches decided after the penalties. We invite the interested reader to check the FIFA website for more details about the rankings' computation.

For our purposes, the FIFA rankings represent a valuable information worthy to be included in our models. For such a reason, we maintained to include these rankings into the usual Poisson scoring intensities specification by multiplying the ranking difference between the two competing teams for a coefficient accounting for its magnitude, similarly as in (5.11).

6.2.4 The DIBP model

To better capture the tournaments' structure just described, we implement a Bayesian diagonal-inflated bivariate-Poisson model with dynamic team-specific offensive and defensive abilities—see Section 5.1.1 in Chapter 5 for further modelling details. Let (Y_{i1}, Y_{i2}) denote as usual the random number of goals scored by the home and the away team in the i-th game, $i = 1, \dots, n$, respectively. The indexes t and k denote the time instant and the team, respectively, whereas "rank" denotes the Coca-Cola FIFA ranking registered just before the two competitions started, at June 1st 2021 for the Euro Cup 2020 and at October 6th, 2022 for the World Cup, respectively. "att" and "def" denote as usual the offensive and the defensive abilities, respectively. The whole model is then given by:

$$
\begin{aligned}
(Y_{i1}, Y_{i2}) &\sim \begin{cases} (1-p)\text{BP}(y_{i1}, y_{i2} | \lambda_{i1}, \lambda_{i2}, \lambda_{i3}) & \text{if } y_{i1} \neq y_{i2} \\ (1-p)\text{BP}(y_{i1}, y_{i2} | \lambda_{i1}, \lambda_{i2}, \lambda_{i3}) + pD(y_{i1}, \eta) & \text{if } y_{i1} = y_{i2}, \end{cases} , \\
\log(\lambda_{i1}) &= \mu + att_{h_i,t} + def_{a_i,t} + \frac{\gamma}{2}(\text{ranking}_{h_i} - \text{rank}_{a_i}) \\
\log(\lambda_{i2}) &= \mu + att_{a_i,t} + def_{h_i,t} - \frac{\gamma}{2}(\text{ranking}_{h_i} - \text{rank}_{a_i}),
\end{aligned}
$$

[3]`https://www.fifa.com/fifa-world-ranking`

$$
\begin{aligned}
\log(\lambda_{i3}) = & \ \rho, \\
att_{k,t} \sim & \ N(att_{k,t-1}, \sigma^2), \\
def_{k,t} \sim & \ N(def_{k,t-1}, \sigma^2), \\
\gamma, \ \rho, \ \mu \sim & \ N(0,1) \\
p \sim & \ \text{Uniform}(0,1) \\
& \sum_{k=1}^{n_t} att_{k,t} = 0, \ \sum_{k=1}^{K} def_{k,t} = 0, \quad k = 1, \dots K, \ t = 1, \dots, T.
\end{aligned} \tag{6.2}
$$

Line (1) in Equation (6.2) displays the likelihood's equations for the diagonal inflated bivariate Poisson (hereafter, DIBP) model; lines (2)-(4) display the log-linear models for the scoring rates $\lambda_{i1}, \lambda_{i2}$ and for the covariance parameter λ_{i3}: note that we assume a constant covariance specification, see (4.17) for further details; lines (5)-(6) display the dynamic prior distributions for the attack and the defence parameters, respectively, as specified in (4.24)—keep in mind that two priors for the first time instant need to be separately specified, as in (4.26), Section 4.4.3; lines (7)-(8) display the prior distributions for the other model parameters; line (9) displays the sum-to-zero identifiability constraints for the team-specific abilities, as in (4.25). Model estimation has been performed through the Hamiltonian Monte Carlo (HMC) sampling—see Section 2.5.1.3—with four chains and 2000 iterations each using the footBayes R package introduced in Chapters 4 and 5. The historical data used to fit the models come from all the international matches played during the previous four years.

6.2.5 Ability estimation

During a Euro/World Cup competition it is plausible that the teams' performances tend to change from the group-stage to the knock-out phase, thus, assessing the evolution of the team-specific abilities is extremely interesting. As broadly explained in the previous chapters, the goal-based models strongly rely on offensive and defensive abilities which need to be estimated according to either a static or a dynamic approach. However, it is not trivial to report a trustful picture of these strengths during a World/Euro cup competition: the main reason is that the use of different training sets could dramatically change and influence these abilities; the second reason is that we usually include a further FIFA ranking covariate to adjust the model fit. Code Snippet 43 reports the main steps for fitting the model until the semifinals of the Euro Cup 2020 and producing the ability estimation plots before the semifinals took place by using the sf foot_abilities function. Note that the argument home_effect is set to FALSE.

Code Snippet 43 Euro Cup 2020: model fit and abilities' estimation.

```
# 1 group-stage
ngames_prev <- 12
euro_data_test <- data.frame(periods = rep(10, ngames_prev),
  home_team = c("Italy", "Wales", "Denmark", "Russia", "England",  "Austria",
               "Netherlands", "Scotland", "Poland",
               "Spain", "Hungary", "Germany"),
  away_team = c("Turkey", "Switzerland", "Finland", "Belgium", "Croatia",
              "FYR Macedonia", "Ukraine", "Czech Republic", "Slovakia",
               "Sweden", "Portugal", "France"),
  home_goals = c(3,1,0,0,1,3,3,0,1,0, 0,0),
  away_goals = c(0,1,1,3,0,1,2,2,2,0,3,1))

# 2 group-stage
euro_data_test2 <- data.frame(periods = rep(11, ngames_prev),
  home_team = c("Russia", "Turkey", "Italy", "Ukraine", "Denmark",
               "Netherlands", "Sweden", "Croatia", "England",
               "Hungary",  "Germany", "Spain"),
  away_team = c("Finland", "Wales", "Switzerland", "FYR Macedonia", "Belgium",
               "Austria", "Slovakia",  "Czech Republic", "Scotland",
               "France", "Portugal", "Poland"),
  home_goals = c(1,0,3,2,1,2,1,1,0,1,4,1),
  away_goals = c(0,2,0,1,2,0,0,1,0,1,2,1))

# 3 group-stage
euro_data_test3 <- data.frame(periods = rep(12, ngames_prev),
  home_team = c( "Italy","Switzerland",  "Ukraine", "Netherlands",
                "Denmark", "Finland",  "Scotland","England",
                "Sweden", "Spain", "Portugal", "Germany" ),
  away_team = c("Wales", "Turkey", "Austria", "FYR Macedonia",
               "Russia", "Belgium",  "Croatia", "Czech Republic",
               "Poland","Slovakia", "France","Hungary"),
  home_goals = c(1,3,0,3,4,0,1,1,3,5,2,2),
  away_goals = c(0,1,1,0,1,2,3,0,2,0,2,2))

# round of 16
ngames_prev <- 8
euro_data_test_round16 <- data.frame(periods = rep(13, ngames_prev),
  home_team = c("Wales", "Italy", "Netherlands", "Belgium",
               "Croatia", "France", "England", "Sweden"),
  away_team = c( "Denmark", "Austria", "Czech Republic", "Portugal",
                "Spain", "Switzerland", "Germany", "Ukraine"),
  home_goals = c(0,2,0,1,3,3,2,1), away_goals = c(4,1,2,0,5,3,0,2))

# quarter of finals
ngames_prev <- 4
euro_data_test_q <- data.frame(periods = rep(14, ngames_prev),
  home_team = c("Switzerland", "Belgium", "Czech Republic", "Ukraine"),
  away_team = c( "Spain", "Italy", "Denmark", "England"),
  home_goals = c(1,1,1,0), away_goals = c(1,2,2,4))
```

Code Snippet 43 *Continue.*

```
# seminfinals
ngames_prev <- 2
euro_data_test_semi <- data.frame(periods = rep(15, ngames_prev),
  home_team = c("Italy", "England"), away_team = c( "Spain",  "Denmark" ),
  home_goals = c(1,2), away_goals = c(1,1))
euro_data <-rbind(euro_data, euro_data_test, euro_data_test2, euro_data_test3,
                  euro_data_test_round16, euro_data_test_q, euro_data_test_semi)

fit_semi <- stan_foot(data = euro_data, model = "diag_infl_biv_pois",
                      home_effect = FALSE, dynamic_type = "seasonal",
                      predict = ngames_prev, ranking = as.data.frame(rankings),
                      cores = 4)

foot_abilities(fit_semi, euro_data,
               teams = c("Denmark", "England", "Italy", "Spain"))
```

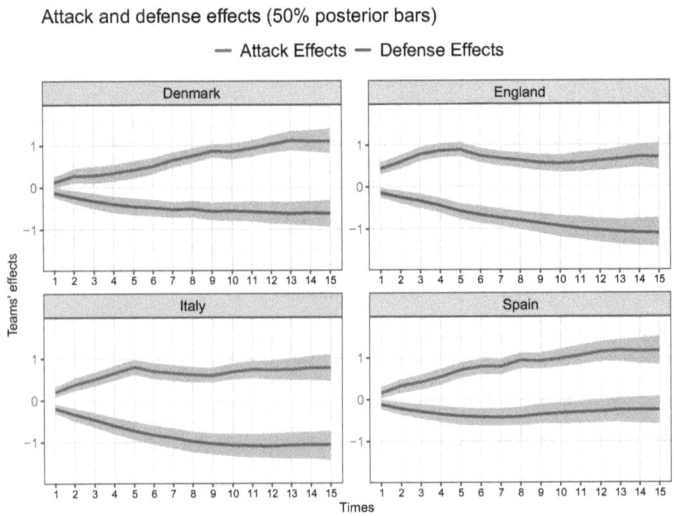

Output 39: Euro Cup 2020: estimated 50% credible intervals (grey ribbons) for the dynamic abilities before the semifinals of the Euro Cup 2020 took place. The red line denotes the posterior median for the attacking ability, whereas the blue line denotes the posterior median for the defensive ability.

We depict in Output 39 the 50% credible intervals for the estimated attacking (red ribbons) and defensive (blue ribbons) abilities for the four teams playing the Euro Cup 2020 semifinals, Denmark, England, Italy, and Spain. As we may notice, the trend is similar for the four teams across the competition: the red line for the attacking strength, the posterior median, increases, whereas the blue line for the defence strength decreases, suggesting how these teams are well progressing during the competition. We remind here that the defence has to be interpreted as the defence weakness, thus the lower is and the better is the estimated defensive performance of the team. Denmark started poorly, by losing against Finland 1-0, but then they dramatically improved their performance, by beating Russia 4-1 in the third match-day of the group-stage and then Wales 4-0 in the round of sixteen. The English team's start has been not totally brilliant, however then they won 2-0 against Germany in the round of sixteen, and defeated Ukraine 4-0 in the quarter of finals. Italy started very well, by beating Turkey and Switzerland 3-0, respectively, but then they stabilized, by equalizing with Spain and England 1-1. Finally, Spain performed very well in the central part of the competition, by scoring five goals against Slovakia first and against Croatia then: however, they stabilized in the quarter of finals and in semifinals, by collecting two draws against Switzerland and Italy, respectively. We need to stress that the four teams exhibit very similar trends; moreover, at the beginning of the tournament the two abilities appear to be very close and approximately equal to zero for each team: this is due to the fact that both the hyper-priors for the first match of the group-stage are Gaussian distributions with mean equal to zero.

6.2.6 Ahead probabilistic predictions

Using training data to obtain ahead probabilistic predictions is pretty easy according to a Bayesian approach: we just need to sample future and observable values from the posterior predictive distributions of the future/held-out matches. For the sake of brevity, we can use the data until time t to fit the model and produce the predictions for $t+1$; in $t+1$ we use the data to make predictions for $t+2$, and so on—we refer to Chapter 2 for a thorough and detailed description of the sampling procedures and the out-of-sample predictions. The computational algorithmic steps we adopted for ahead predictions in international competitions are proposed in Algorithm 10.

Algorithm 10 Ahead probabilistic predictions: international matches.

At time t:

STEP 1 : use training data comprehensive of the qualifiers, friendly, and Nation's league matches constituting the training set. If $t > 0$, use also the Qatar 2022 results up to match-day $t-1$

STEP 2 : estimate the DIBP model (6.2) through the footBayes package and obtain the posterior estimates for the model's parameters

STEP 3 : obtain probabilistic predictions for time $t+1$ from the posterior predictive distribution, $f(\tilde{\mathcal{D}}|\mathcal{D})$

STEP 4 : create the new training set by embedding matches up to match-day t and set $t \leftarrow t+1$ and go back to STEP 1.

Code Snippet 44 summarizes the steps for producing probabilistic predictions for the third match-day from the group-stage of the Qatar World Cup 2022. In Table 6.1 we provide some probabilistic predictions for the third match-day from the group-stage of the Qatar World Cup 2022: the third, fourth and fifth columns report the posterior predictive probabilities for the home win, the draw, and the away win, respectively, whereas the sixth column reports the "most likely outcome" (MLO) with the associated probability—note that in the international competitions such as Euro and World Cups the terms "home" and "away" do not have a proper meaning, however we keep these definitions for consistency with the previous chapters; the home effect here is not considered in the model. However, the FIFA institution always establishes, just as a formal convenience, the home and the away team for each match in the international matches. Note that the same information contained in Table 6.1 can be graphically depicted by the Output 40, a *chessboard* plot produced by the foot_prob function where darker (lighter) regions correspond to higher (lower) probabilities, and the red square denotes the actual final result. In this plot the first team name in the single labels denotes the "favourite" team, whereas the second name denotes the "underdog" team—the term "underdog" denotes the team associated with lower winning chances; moreover, we depict the most balanced matches to the most apparently unbalanced ones from the left top corner to the right bottom corner—note that in these plots, unlike for what happens in the Table 6.1, a sort of results' "truncation" occurred, since the highest number of goals considered for both the teams is four. Thus, the match Ecuador-Senegal (top left corner in the plot) is the most balanced under the DIBP model—Ecuador and Senegal have both a 35% chance of winning, and the most likely exact outcome is 0-0, with about 16% of chance—whereas Brazil-Cameroon (bottom right corner in the plot) is the most unbalanced, given that Brazil has a 66% winning chance against the 8% for Cameroon: the final actual results for these two matches were 1-2 and 0-1, thus in both cases the underdog team won the match. In the first case, the result 1-2 had about the 6% probability to occur, whereas the result 0-1 for Cameroon had about the 1% chance to occur. We could note that many MLOs in Table 6.1 are 0-0, and this is a direct implication of the diagonal inflation of the model.

The same kind of representation is plotted in Output 41 for the two Qatar 2022 finals, Argentina-Francia—the final actual result within the regular 90 minutes was 3-3—and Morocco-Croatia—for which the final actual result was 1-2—respectively, whereas the corresponding numerical results are reported in Table 6.2. Note that the whole R code for reproducing the results is not shown here for sake of brevity. Argentina was given a winning chance of 41% within the 90 minutes, whereas Croatia was given a winning chance of 27%.

Code Snippet 44 World Cup 2022: model fit and probabilistic predictions.

```
# 1 group-stage
ngames_matchday1 <- 16
wc_data_train_matchday_1 <- data.frame(periods =
  rep(length(unique(wc_data_train$periods))+1, ngames_matchday1),
  home_team = c("Qatar", "England" , "Senegal",
                "United States", "Argentina", "Denmark",
                "Mexico", "France", "Morocco", "Germany",
                "Spain","Belgium", "Switzerland","Uruguay",
                "Portugal","Brazil"),
  away_team = c("Ecuador", "Iran", "Netherlands",
                "Wales", "Saudi Arabia", "Tunisia",
                "Poland", "Australia", "Croatia",
                "Japan", "Costa Rica","Canada",
                "Cameroon","South Korea",
                "Ghana","Serbia"),

  home_goals = c(0,6,0,1,1,0,0,4,0,1,7,1,1,0,3,2),
  away_goals = c(2,2,2,1,2,0,0,1,0,2,0,0,0,0,2,0),
  tournament = rep("World Cup 2022", ngames_matchday1 ))

# 2 group-stage
ngames_matchday2 <- 16
wc_data_train_matchday2 <- data.frame(periods =
  rep(length(unique(wc_data_train$periods))+2, ngames_matchday2),
  home_team = c("Wales","Qatar","Netherlands",
                "England", "Tunisia","Poland",
                "France","Argentina", "Japan","Germany",
                "Belgium","Croatia","Cameroon","Brazil",
                "Portugal","South Korea"),
  away_team = c( "Iran",  "Senegal","Ecuador",
                 "United States", "Australia",
                 "Saudi Arabia",
                 "Denmark","Mexico","Costa Rica","Spain",
                 "Morocco","Canada","Serbia","Switzerland",
                 "Uruguay","Ghana"),
  home_goals = c(0,1,1,0,0,2,2,2,0,1,0,4,3,1,2,2),
  away_goals = c(2,3,1,0,1,0,1,0,1,1,2,1,3,0,0,3),
  tournament = rep("World Cup 2022",ngames_matchday2))

# 3 group-stage
ngames_matchday3 <- 16
wc_data_train_matchday3 <- data.frame( periods =
  rep(length(unique(wc_data_train$periods))+3, ngames_matchday3),
  home_team = c("Ecuador", "Netherlands",  "Iran", "Wales",
                "Tunisia", "Australia", "Poland", "Saudi Arabia",
                "Croatia", "Canada", "Japan", "Costa Rica",
                "South Korea", "Ghana", "Serbia","Cameroon"),
  away_team = c( "Senegal", "Qatar", "United States",  "England",
                 "France", "Denmark", "Argentina", "Mexico",
                 "Belgium", "Morocco", "Spain",  "Germany",
                 "Portugal", "Uruguay", "Switzerland", "Brazil"),
  home_goals = c(1,2,0,0,1,1,0,1,0,1,2,2,2,0,2,1),
  away_goals = c(2,0,1,3,0,0,2,2,0,2,1,4,1,2,3,0),
  tournament = rep("World Cup 2022", ngames_matchday3))
```

Code Snippet 44 *Continue.*

```
wc_data_stan <-rbind(wc_data_train, wc_data_train_matchday_1,
                     wc_data_train_matchday2, wc_data_train_matchday3)

fit_group3 <- stan_foot(wc_data_stan[,-6], model = "diag_infl_biv_pois",
                        home_effect = FALSE, dynamic_type = "seasonal",
                        predict = ngames_matchday3,
                        ranking = as.data.frame(rankings), cores = 4)
foot_prob(data = wc_data_stan[,-6], object = fit_group3)
```

TABLE 6.1
Qatar World Cup 2022, third match-day of the group-stages: probabilistic predictions from the posterior predictive distribution. "MLO" denotes the most likely outcome

Home team	Away team	Home win	Draw	Away win	MLO
Ecuador	Senegal	0.35	0.30	0.35	0-0 (0.16)
Netherlands	Qatar	0.71	0.19	0.10	1-0 (0.09)
Iran	United States	0.35	0.28	0.37	0-0 (0.12)
Wales	England	0.19	0.24	0.57	0-1 (0.11)
Tunisia	France	0.19	0.26	0.55	0-0 (0.13)
Australia	Denmark	0.30	0.29	0.41	0-0 (0.15)
Poland	Argentina	0.19	0.28	0.53	0-0 (0.15)
Saudi Arabia	Mexico	0.33	0.35	0.32	0-0 (0.23)
Croatia	Belgium	0.43	0.26	0.31	0-0 (0.10)
Canada	Morocco	0.20	0.28	0.52	0-0 (0.15)
Japan	Spain	0.21	0.24	0.54	0-1 (0.11)
Costa Rica	Germany	0.21	0.27	0.52	0-0 (0.13)
South Korea	Portugal	0.19	0.23	0.58	0-1 (0.11)
Ghana	Uruguay	0.26	0.25	0.49	0-1 (0.11)
Serbia	Switzerland	0.35	0.30	0.35	0-0 (0.16)
Cameroon	Brazil	0.08	0.26	0.66	0-1 (0.19)

6.2.7 Winning probabilities

One of the most intriguing tasks sports data scientists are asked during a knock-out tournament is to estimate the final winning probabilities as the tournament evolves. We should notice that obtaining this kind of predictions is slightly more complicated than predicting at time t the results for the match-day $t+1$: in fact, we need to generate some future scenarios and draw the knock-out stage from the current stage until the end of the tournament: then, we just need to count how many times the team k wins the tournament across the distinct scenarios. As one can notice, this can be computationally expensive, as shown and explained in Algorithm 11.

Algorithm 11 Winning probabilistic predictions

At time t:

Step 1 : use training data comprehensive of the qualifiers, friendly, and Nation's league matches constituting the training set. If $t > 0$, use also the tournament results up to match-day $t - 1$

Step 2 : estimate the DIBP model (6.2) through the footBayes package and obtain the posterior estimates for the model's parameters

Step 3 : construct M predictive scenarios for the upcoming tournament phases from the posterior predictive distribution, and count how many times a given team wins the final tournament, using $q_k = \#\{\text{team } k \text{ wins}\}/M$ as its estimated winning probability.

TABLE 6.2

Qatar World Cup 2022, finals: probabilistic predictions from the posterior predictive distribution. "MLO" denotes the most likely outcome

Home team	Away team	Home win	Draw	Away win	MLO
Argentina	France	0.410	0.307	0.283	0-0 (0.172)
Croatia	Morocco	0.272	0.335	0.393	0-0 (0.211)

Once the tournament has reached the semifinals, we estimated the winning probabilities for the Euro Cup 2020 and the World Cup 2022 in Table 6.3: as it may be noted, the two winner teams, Italy and Argentina, got the highest winning chances before the semifinals took place, 34% and 45%, respectively. Although this is not to be taken as a sort of "gold oracle", we strongly believe that our simple DIBP model can capture well some dynamics underlying the knock-out tournament evolution. We may in fact also appreciate that for both the tournaments the final actual ranking—Italy and England for Euro 2020, for which there is not the third-place final; Argentina, France, Croatia and Morocco for the World Cup—exactly mirrors the probabilistic ranking implied by the model's winning probabilities reported in the third column in the table.

6.2.8 Expected goals

Nowadays expected goals (Rathke, 2017)—xG, hereafter—represent an intriguing measure that seeks to account for the offensive potential produced by a team in a given game or the potential scoring chances of a single player. The number of xG translates the goals that one team or player would have expected to score. In order to quantify the xG in a given match, we would need to construct suitable algorithms/models and use some influential covariates/features, such as the distance to goal, the angle of shot, the body part, etc.

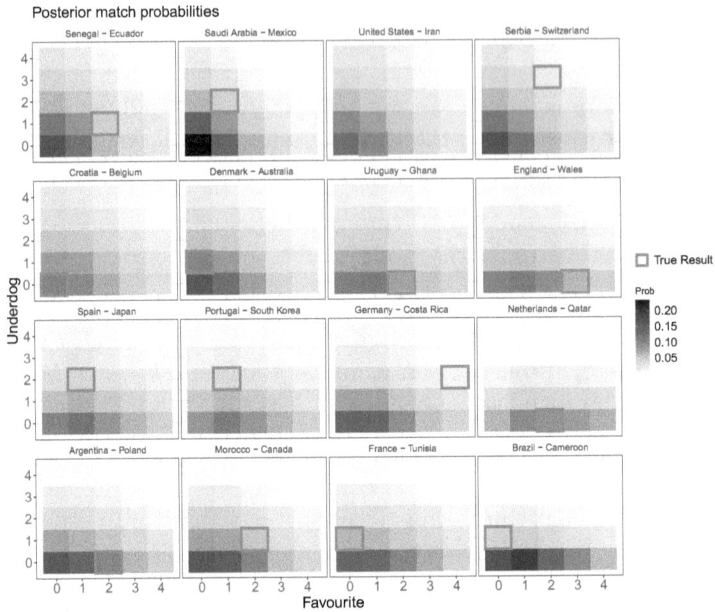

Output 40: Qatar World Cup 2022, third match-day of the group-stages: probabilistic predictions from the posterior predictive distribution. The first team listed in each sub-title is the "favourite" (x-axis), whereas the second team is the "underdog" (y-axis). The 2-way grid displays the 16 held-out matches in such a way that closer matches appear at the top-left of the grid, whereas more unbalanced matches ("blowouts") appear at the bottom-right. The matches are then ordered from top-left to bottom-right in terms of increasing winning probability for the favourite teams. Darker regions correspond to more likely results, whereas red squares denote the actual observed results.

Although xG represent nowadays a very hot topic in sport newspapers and specialized magazines, we need to note that to construct suitable xG models one would need rich, granular, and sophisticated data, which are usually far to be accessible to a large audience: among them, one could consider the location of every single match shot; how the ball was delivered to the person making the shot; whether the shot was made by foot or by header, and so on. However, this kind of modelling procedure goes beyond the scopes of this book. Moreover, there is not a gold-standard scientific/statistical modelling protocol for xG yet, and this makes these measures quite subjective and, at least now, hardly reproducible.

During the two knock-out tournaments we decided to report a measure of xG simply based on a point estimate of the estimated parameters from the DIBP model (6.2), by computing the xG for team A and team B respectively

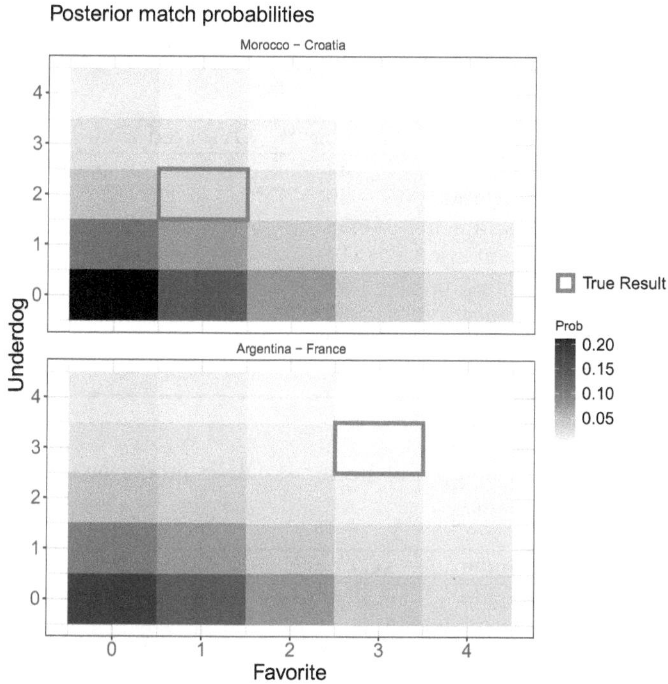

Output 41: Qatar World Cup 2022, finals: probabilistic predictions from the posterior predictive distribution. "favourite" and "underdog" denote the favourite and the underdog team, respectively. The first team listed in each sub-title is the "favourite" (x-axis), whereas the second team is the "underdog" (y-axis). Darker regions correspond to more likely results, whereas red squares denote the actual observed results.

as:

$$\begin{aligned} \text{xG}_A &= \exp\{\widehat{att}_A + \widehat{def}_B + \frac{\hat{\gamma}}{2}(\text{ranking}_A - \text{rank}_B)\} \\ \text{xG}_B &= \exp\{\widehat{att}_B + \widehat{def}_A - \frac{\hat{\gamma}}{2}(\text{ranking}_A - \text{rank}_B)\}, \end{aligned} \tag{6.3}$$

where $\widehat{att}$ and $\widehat{def}$ denote the posterior medians for the attacking and the defensive parameters, respectively. Table 6.4 reports the xG—third and fourth column—computed for the two semifinals of the Euro Cup 2020, Italy against Spain and England against Denmark, respectively. The table also reports the actual observed goals—fifth and sixth column. We could notice that both the matches were predicted to be very tight in terms of the final number of expected scores within the regular times, and the global predicted balance was mirrored by the observed results.

TABLE 6.3
Euro Cup 2020 and World Cup 2022 estimated winning probabilities before the semifinals took place

	Team	Winning %	Observed rank
Euro Cup 2020	Italy	34	1
	England	30	2
	Spain	22	not assigned
	Denmark	14	not assigned
World Cup 2022	Argentina	45	1
	France	35	2
	Croatia	12	3
	Morocco	8	4

TABLE 6.4
Euro Cup 2020 semifinals: xG computed from the DIBP model against the number of actually observed scores

favourite	underdog	xG_f	xG_u	y_1	y_2
Italy	Spain	1.36	0.85	1	1
England	Denmark	1.41	0.81	2	1

6.2.9 What happened, what we predicted

The task of predicting a knock-out tournament is not trivial and we should be always aware of the "risks" accompanying this action. In fact, better teams are usually quite favourite against weaker teams, nonetheless some "underdogs" are sometimes able to overturn the forecast. For instance, consider the match Argentina-Saudi Arabia, played during the first match-day of the Qatar World Cup 2022 group-stage, group C. According to the majority of the bookmakers, the chance of win for Saudi Arabia did not exceed 4-5%, however our DIBP model assigned to Saudi Arabia a winning chance approximately equal to 6%. Argentina immediately scored with Leo Messi, however Saudi Arabia overturned the result in the second half and defeated Argentina 2-1, by getting a very surprising result. As another example of an underdog beating the favourite, we could also consider the match Germany-Japan, group-stage E, where Japan defeated Germany 2-1 in the last minutes of the match.

Although the football matches, in particular those arising in the knock-out tournaments, are difficult to be forecasted, we strongly maintain that our model has been able to provide satisfactory predictive results. If we look at the global pseudo-R^2 for the probabilities of the process win/draw/loss this is equal to 0.356 for the Qatar World Cup 2022 and 0.377 for the Euro 2020, whereas the Average of Correct Probabilities (ACP) in (3.14) in Section 3.4.5 is equal to 0.41 for the Qatar World Cup 2022 and 0.424 for the Euro 2020. We stress the fact that a "random" classifier would obtain 1/3 in terms of both the metrics: the results above clearly show that adopting a statistical model in place of a random classifier has some relevant advantages.

Moreover, another predictive confirmatory tool for our model comes from the winning probabilities reported in Table 6.3: for both the knock-out tournaments we have been somehow able to predict in advance the favourite final winner and, in general, the final rankings implied by the model's probabilities.

As a final predictive tool, the users could compute some of the model's predictive measures reported in Chapter 3, Section 3.4, such as the Brier score, the ACP, and the pseudo-R^2, by using the compare_foot function of the footBayes package, as reported in Code Snippet 45 and Output 42—here just for the third match-day of the group-stage used as held-out data.

Code Snippet 45 World Cup 2022: predictive performance of the DIBP model.

```
compare_foot(fit_group3, test_data = wc_data_train_matchday3)
```

```
Predictive Performance Metrics
 Model   RPS accuracy brier pseudoR2   ACP
   fit 0.278    0.500 0.644    0.332 0.393
```

Output 42: World Cup 2022: predictive performances of the DIBP model.

6.3 Comparison with odds forecasters

It is not easy to compare in detail the predictive performance of our model in the two knock-out tournaments with those provided by the bookmarkers. There are many reasons for this lack of comparisons. First of all, there is a debate about which betting odds should be more representative for a given match: as far as we know, the bookies adjust their odds before and during the match itself, then they provide a sort of dynamic odds trend, and for such a reason it is not clear at all when considering the most plausible set of odds. Moreover, they usually adjust the odds on the ground of the bettors' market, and it is implicit that, especially a few hours before the match begins, some adjustments occur only due to economic factors and bettors' fluctuations.

As far as we know there is not even a public historical repository for the odds provided by the bookies for international matches, which makes the process of comparison quite hard. For both the knock-out tournaments, we considered the betting odds provided one hour before the matches from some Italian and European bookmakers: in the majority of the cases, the Brier score (3.9) in Section 3.4.3 and the pseudo-R^2 (3.15) in Section 3.4.6 values obtained from the DIBP model were globally better than those obtained from the bookmakers' odds.

Finally, even if a whole and thorough comparison with bookmakers and odds forecasters goes out from the purposes of this book, we claim that our model has been so far able to produce overall good predictive performances for the Euro Cup 2020 and the World Cup 2022. In Chapter 7 we provide a whole betting simulation to compare the probabilistic performances given by our models and the bookmarkers.

6.4 Future research

As previously remarked, the main purpose of this chapter is to give the reader an overview of our modelling and predictive experience during the last Euro and World Cups. We strongly feel that many of these analyses could be enriched, updated, if not improved, and for these reasons we invite the interested reader to try on their own to do better, by using the reproducible R code and the data accompanying this book.

Thus, we conclude this modelling journey by collecting some further points left for future research and interest:

- weight the training set matches according to their importance;
- evaluate the inclusions of other types of rankings and other covariates at team-level and match-level;
- include some information about the rosters/players' participation;
- train distinct models and possibly obtain some predictions through Bayesian model averaging techniques;
- update the final winning probabilities as long as the tournament evolves;
- produce online predictions as the matches evolve.

6.5 Summary and closing remarks of Chapter 6

In this chapter we provided a thorough and detailed overview of our modelling experience for two of the recent most important football international tournaments, the Euro Cup 2020—actually played in 2021 due to the Covid-19 outbreak—-and the World Cup 2022 hosted by Qatar by using the footBayes package.

Some modelling strategies for these kinds of tournaments have been proposed in the previous chapters, however we provide in Section 6.2.1 a deep focus on data acquisition and some choices left to the user choice. The two competitions exhibit a similar structure, as explained in Section 6.2.2, with a partial round-robin scheme followed by a knock-out phase. The inclusion of the FIFA rankings is proposed in Section 6.2.3, whereas the diagonal-inflated bivariate Poisson model chosen for the analysis is thoroughly presented in Section 6.2.4. The team-specific abilities are plotted and described in Section 6.2.5, whereas out-of-sample predictions represent the focus of Sections 6.2.6 and 6.2.7 in terms of ahead predictions and winning probabilities. A naive use of the so-called expected goals, or xG, is the focus of Section 6.2.8, whereas a final analysis of predictive accuracy is given in Section 6.2.9. Section 6.3 quickly focuses on a comparison with the bookmakers, whereas some points of improvement and future research are provided in Section 6.4.

7

Compare statistical models' performance with the bookmakers

7.1 How odds relate to probabilities

As widely known, there is a strong connection existing between betting odds and probabilities, broadly investigated over the last decades. However, there is often a general lack of awareness in the odds' interpretation and in their final derivation. To begin with some intuitive arguments, the odd of a given event, say 2.5, is usually specified as the amount of money we would win if we bet one unit on that event: thus, if we bet 1 Euro on the event E and then the event E is actually observed, we obtain a cash-flow of 2.5 Euro. To get a realistic measure of the likelihood of a given event in terms of the bookmakers' evaluation, we can compute the *inverse odd*, usually denoted with 1:2.5: however, the latter is a non-coherent probability associated to that event. In fact, summing the betting odds associated to the sequence of possible occurrences for a given event—-as the home win, the draw, and the away win for a football match—does not yield 1, rather the sum of the inverse odds is greater than one (Dixon and Coles, 1997) to allow the bookmakers to make their profit, or margin. We will provide some examples in the remainder of the chapter.

We start by introducing some convenient notation for the rest of the chapter. We denote with:

$$
\begin{aligned}
\boldsymbol{o}_i &= \{o_i^{Home}, o_i^{Draw}, o_i^{Away}\}, \quad i = 1, \dots, n \\
\boldsymbol{\pi}_i &= (\pi_i^{Home}, \pi_i^{Draw}, \pi_i^{Away}), \quad i = 1, \dots, n \\
\boldsymbol{p}_i &= (p_i^{Home}, p_i^{Draw}, p_i^{Away}), \quad i = 1, \dots, n \\
\Delta_i &= \{\text{“}\mathit{Home\ win}\text{”}, \text{“}\mathit{Draw}\text{”}, \text{“}\mathit{Away\ win}\text{”}\}, \quad i = 1, \dots, n,
\end{aligned}
$$

the vector of the inverse betting odds $\boldsymbol{o}_i$, the vector of the estimated betting probabilities from the bookmaker(s) $\boldsymbol{\pi}_i$, the vector of estimated probabilities under the model $\boldsymbol{p}_i$, and the set of the three-way results Δ_i for the i-th game, respectively.

Even though they cannot be directly used as probabilistic objects, there is some empirical evidence that the betting odds are the most accurate available source of probability forecasts in many sports (Štrumbelj, 2014); in other words, predictions based on odds-probabilities have been shown to be better,

DOI: 10.1201/9781003186496-7

or at least as good as, statistical models which use sport-specific predictors and/or expert tipsters.

However, the betting odds could be easily transformed into coherent probabilities through some simple mathematical manipulation. There is a strong debate over which method to use for inferring a set of probabilities from the raw betting odds, the two main procedures proposed in the literature are: the *basic normalization*, which consists in dividing the inverse odds by the booksum, i.e. the sum of the inverse betting odds, as broadly explained in Štrumbelj (2014); and the *Shin's procedure* described in Shin (1991, 1993). Štrumbelj (2014), Cain et al. (2002, 2003), and Smith et al. (2009) show that Shin's probabilities improve on the basic normalization: in Štrumbelj (2014) this result has been achieved by the application of the Ranked Probability Score (RPS) (Epstein, 1969) and the Brier score (Brier et al., 1950), two well-known discrepancy measures between the probability of a three-way process outcome and the actual outcome, as explained in Sections 3.4.3 and 3.4.4 in Chapter 3. We give here a brief overview of the two transformation methods.

7.1.1 Basic normalization

As explained by Dixon and Coles (1997), one could easily infer a vector of probabilities from the original betting odds just by using the following normalization rule:

$$\pi_i^j = \frac{o_i^j}{\beta}, \; i = 1, \ldots, n, \; j \in \Delta_i, \tag{7.1}$$

where $\beta = \sum_j o_i^j$ is the so called *booksum* (Štrumbelj, 2014). The method has gained a great popularity due to its simplicity and is usually adopted as a benchmark for probabilistic forecasts derived from the bookmakers odds.

7.1.2 Shin's procedure

In the model proposed by Shin (1993), the financial market is populated by the *market makers* and the *traders*. In the specific setting where the market is the market for bets, the bookmakers are the marked makers who specify their odds with the aim to maximise their expected profit, whereas the traders are represented by the *uninformed bettors*. However, the novelty introduced by Shin is to augment this market by considering a third actor, generically represented by the *insider traders*. An insider trader is a particular actor who, due to his/her superior information, is assumed to *already* know or partially know the outcome of a given event—e.g. a football match, a horse race, etc.—before that specific event takes place. Their contribution in the global betting volume is quantified by the percentage z. Assume that the bookmaker has probabilistic beliefs expressed by $\boldsymbol{\pi}_i$ for the event i, the total expected profit for the bookmaker in the match i is then given by:

$$G(\pi) = 1 - \sum_{j \in \Delta_i} \pi^j o^j (\pi^j (1 - z) + z), \tag{7.2}$$

where we suppressed the upper index i for easing the notation, and Δ_i contains all the possible event's occurrences. The bookmaker sets the vector of inverse odds $\boldsymbol{o}$ in order to maximize the expected profit, subject to the constraints: $0 \leq o^j \leq 1$, for each occurrence j.

Jullien et al. (1994) explicitly derived a closed-form expression for the betting probabilities π^j, depending on z, derived from the inverse betting odds o^j under the Shin's approach as:

$$\pi^j(z) = \frac{\sqrt{z^2 + 4(1-z)(o^j)^2/\beta} - z}{2(1-z)}, \; j \in \Delta_i, \tag{7.3}$$

referred in the literature as *Shin's probabilities*. Equation (7.3) is a function depending on the insider trading rate z: in order to estimate it, Jullien et al. (1994) suggest to use non-linear least squares as:

$$\hat{z} = \underset{z}{\text{argmin}} \left[\sum_{j \in \Delta_i} \pi^j(z) - 1 \right]^2,$$

which gives the closed-form solution obtained by Štrumbelj (2014)

$$z = \sum_{j \in \Delta_i} \sqrt{z^2 + 4(1-z)(o^j)^2/\beta} - 2, \tag{7.4}$$

by using fixed-point iteration. Note that as the booksum

approaches 1, the proportion of insider traders z goes to 0, reducing Shin's approach to basic normalization. $\hat{z}$ may be defined as the minimum rate of insider traders that yields those probabilities corresponding to the vector of inverse betting odds $\mathbf{o}$.

As it is usually remarked, for instance in Egidi et al. (2018b) and in Figure 7.1 for the English Premier League from 2007 to 2017, the draw probabilities obtained under the basic normalization tend to be slightly higher than those obtained under Shin's procedure: in such a sense, basic normalization implements a *uniform* adjustment of the three probabilities, whereas the Shin's procedure addresses the transformation by assuming a non-uniform approach. Shin (1993) and Štrumbelj (2014) remark that the Shin's normalization improves on basic normalization; however, there are not mathematical proofs about the supposed supremacy of one method over another, rather the assessment should be performed via some measures and indicators of probabilistic forecasting such as the Ranked Probability score (Epstein, 1969), or the Brier score described in Section 3.4.3 in Chapter 3. We stress that a thorough and detailed motivation behind the Shin's probabilities formulation goes beyond the purposes of this chapter and the whole book in general. We are rather interested on a broad overview about probabilistic forecasts derived from the bookmakers. We refer the interested reader to the work of Shin (1991, 1993).

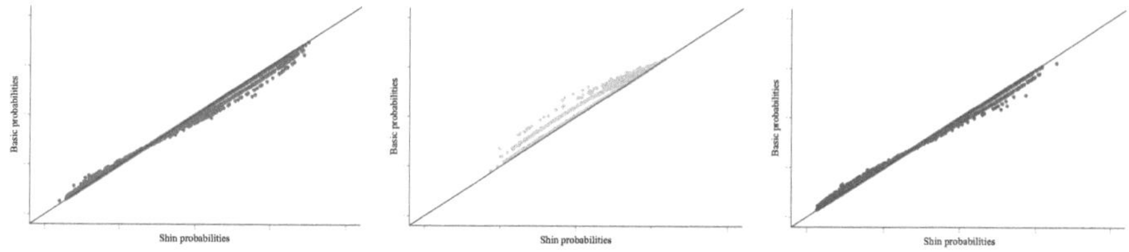

(a) $\rho = 0.99$, lOR = -0.01 (b) $\rho = 0.92$, lOR = 0.02 (c) $\rho = 0.99$, lOR = 0.002

FIGURE 7.1
From Egidi et al. (2018b): comparison between home win (a), draw (b) and away win (c) Shin probabilities (x-axis) and the basic normalized probabilities (y-axis) for the English Premier League Seasons from 2007–2008 to 2016–2017, according to seven different bookmakers. For each three-way outcome, ρ is the Pearson's correlation coefficient and lOR a global log-odds ratio between basic and Shin probabilities over all the matches and all the different bookmakers.

7.1.3 Regression analysis

An alternative way to derive probabilities from betting odds is to use a regression analysis approach to predict the outcome probabilities from a set of bookmakers' odds. As remarked by Štrumbelj (2014), this approach requires a historical set of betting odds and match outcomes, which can be used to estimate the parameters of the model. For sports with three outcomes, as in football, an ordered logistic regression model is used by Train (2009) in their Chapter 7, whereas for two outcomes a logistic/probit regression can be safely adopted (Train, 2009, Chapter 3).

7.2 The bookmaker market: Expected profit, fairness, and margin

As it is widely known and stressed in the previous section, inverse betting odds do not correspond to coherent probabilities, unless some transformations and normalizations as those introduced in the previous sections are applied. This is due to the fact that the bookmakers inflate their odds by allowing for a gain margin: as a consequence, when buying a given event at a given price, the bet will be always *not favourable* from the bettor's perspective. As explained by Shin (1993), if the prices were proportional to true probabilities, the winnings ratio would be constant over all the subjects. However, the odds are biased in such a way that contingent claims on the favourites are cheaper (relative to true probabilities) than the claims on the longshots. This is the so-called *favourite-longshot* bias.

In what follows, we mathematically describe the bookmaker and the bettor positions, respectively, by assuming different scenarios for the underlying betting market where they act. To better understand how the two distinct actors—bookmakers and bettors—achieve their respective aims, we will need to think in terms of *expected profits* (de Finetti, 1931; De Finetti, 1970). Perhaps, in the next sections we analyze and introduce the concepts of profit and expected profit for both the *market's actors*, the *bettor* and the *bookmaker*, respectively. In doing this, we will assume some comfortable operational assumptions, such as considering an artificial market populated only by one bettor. However, one could use the simple mathematics therein introduced to make these situations less artificial, and, for instance, study and infer some more complicated behaviours of the betting companies with regard to a population of bettors. This and other purposes, despite very interesting, go out from the scopes of this book.

7.2.1 Market with one bookmaker and one event

As a starting point, consider a betting market populated by only one bookmaker and one bettor, for simplicity. The bookmaker holds an event E associated with a true probability p to occur, and quote it as $1/o$, where o represents the inverse odd. Taking the bettor's perspective, he could consider to bet an amount S on the event E, quoted as $1{:}1/o$, either winning then the amount $W = S/o$ if the event E occurs, with probability p, or lose S if the event E does not occur, with probability $1-p$. We denote the bettor's profit with G_{be}, and the binary event realization with the couple $(E, \bar{E})$. The event profit for the bettor is then represented by:

$$\begin{aligned} G_{be}(E) &= (W - S) \times p \\ G_{be}(\bar{E}) &= \ - S \times (1-p), \end{aligned} \tag{7.5}$$

whereas from the bookmaker's viewpoint the profit G_{bo} is given by the opposite situation:

$$\begin{aligned} G_{bo}(E) &= (S - W) \times p \\ G_{bo}(\bar{E}) &= S \times (1-p). \end{aligned} \tag{7.6}$$

We compute now the *expected profit* for the event E, by taking both the positions:

$$\begin{aligned} \mathrm{E}(G_{be}) &= (W-S)\times p - S\times(1-p) = Wp - S = S\left(\frac{p}{o}-1\right) \\ \mathrm{E}(G_{bo}) &= (S-W)\times p + S\times(1-p) = S - Wp = S\left(1-\frac{p}{o}\right). \end{aligned} \tag{7.7}$$

We easily realize how the expected profit for the two positions strictly depends on the invested amount S and the probability-odds ratio p/o: if the probability of E is higher than the bookmaker inverse odd—we remind, the inverse odd is a non-coherent probability evaluation—being then $p/o > 1$, the bettor yields a positive expected profit; conversely, the bookmaker yields a positive expected profit if and only if $p/o < 1$. In rough words, the bettor should invest the money on the event E if and only if the probability of E exceeds the bookmakers' inverse odd for E. However, in a horse race or in a football match, the difficulty of obtaining favourable expected profits for the bettors is due to the ignorance about the true occurrence probabilities for the events of interest.

At this step the arguments in favour of an eventual positive expected return invite us to consider the notion of *equity*, or *fairness*, introduced by the Italian mathematician Bruno de Finetti in De Finetti (1970): the bet above is fair if and only if the expected profit is zero for both the positions, meaning that the bet will not give any advantage to either the bettor or the bookmaker. Mathematically speaking, this means:

$$\begin{aligned} \mathrm{E}(G_{be}) = & \; \mathrm{E}(G_{bo}) = 0 \Leftrightarrow \\ Wp - S = & \; S - Wp = 0 \Leftrightarrow \\ W = & \; \frac{S}{p} \Leftrightarrow \\ S/o = & \; S/p \Leftrightarrow \\ o = & \; p, \end{aligned} \tag{7.8}$$

which means that in order for a bet to be fair *the inverse odd of the bookmaker must coincide with the true probability of occurrence p for the event E*: in other words, the odds formulated by the bookmaker should exactly mirror then the true event probability.

The bet is instead *unfair* if the expected profit is in favour either of the bookmaker or the bettor, and this applies whenever $\mathrm{E}(G) \neq 0$. For instance, a positive expected profit for the bettor implies $\mathrm{E}(G_{be}) > 0 \Leftrightarrow S/o > S/p$, then $o < p$: it is then immediate to notice that managing the inverse odd o affects the fairness of the bet. Suppose in fact we could assume a new inverse odd o', with $o' > p$: this implies a winning amount $W' = S/o' < S/p = W$, meaning that $\mathrm{E}(G_{bo}) > 0$, in such a way the bet would be always favouring the bookmaker. Then, increasing/decreasing the inverse odd makes the bet unfair, favouring the bettor or the bookmaker: in general, all the odds in the market are biased in favour of the bookmakers, and this represents the *favourite-longshot bias* for the betting companies.

7.2.2 Market with one bookmaker and more events

We can generalize the setting above, by using a collection of mutually exclusive events $E_1, E_2, \ldots, E_k$, such that $\Pr(\cup_{i=1}^k E_i) = \sum_{i=1}^k \Pr(E_i) = 1$, and $E_i \cap E_j = \emptyset$, when $i \neq j$. A typical example could be given by a horse race where each E_i represents the event: "*the horse i wins the race*". We further assume that each event is associated with some "true" occurrence probabilities $p_1, p_2, \ldots, p_k$, whereas the inverse odds provided by

TABLE 7.1
Market with one bookmaker and k mutually exclusive events

event	true p	inv. odds	prices	$\mathrm{E}(G_{be})$	$\mathrm{E}(G_{bo})$
E_1	p_1	o_1	$-S_1$	$S_1\frac{p_1}{o_1}-\sum_i S_i$	$\sum_i S_i - S_1\frac{p_1}{o_1}$
E_2	p_2	o_2	$-S_2$	$S_2\frac{p_2}{o_2}-\sum_i S_i$	$\sum_i S_i - S_2\frac{p_2}{o_2}$
...	...	...	...	...	...
E_i	p_i	o_i	$-S_i$	$S_i\frac{p_i}{o_i}-\sum_i S_i$	$\sum_i S_i - S_i\frac{p_i}{o_i}$
...	...	...	...	...	...
E_k	p_k	o_k	$-S_k$	$S_k\frac{p_k}{o_k}-\sum_i S_i$	$\sum_i S_i - S_k\frac{p_k}{o_k}$

the
bookmaker operating in the market are given by $o_1, o_2, \ldots, o_k$. Moreover, we assume that the bettor pays the prices $S_1, S_2, \ldots, S_k$ to win one among the amounts $W_1 = S_1/o_1, W_2 = S_2/o_2, \ldots, W_k = S_k/o_k$ in case one of the mutual events takes place. See Table 7.1 for a detailed illustration of the bettor and bookmaker positions. From the bettor's perspective, the profit for the couple $(E_i, \bar{E}_i)$ takes places is:

$$\begin{aligned} G_{be}(E_i) &= (W_i - \sum_{i=1}^{k} S_i) \times p_i \\ G_{be}(\bar{E}_i) &= -(1-p_i) \times \sum_{i=1}^{k} S_i, \end{aligned} \tag{7.9}$$

whereas symmetrically the profit for the bookmaker is:

$$\begin{aligned} G_{bo}(E_i) &= \left(\sum_{i=1}^{k} S_i - W_i\right) \times p_i \\ G_{bo}(\bar{E}_i) &= (1-p_i) \times \sum_{i=1}^{k} S_i. \end{aligned} \tag{7.10}$$

From the equations above, we may then define the expected profits for the two actors with some simple algebraic manipulations:

$$\begin{aligned} \mathrm{E}(G_{be}) &= \sum_{i=1}^{k} \left[S_i \frac{p_i}{o_i} - \sum_i S_i \right] \\ \mathrm{E}(G_{bo}) &= \sum_{i=1}^{k} \left[\sum_i S_i - S_i \frac{p_i}{o_i} \right]. \end{aligned}$$

Similarly as what happens for the one-event market presented in Section 7.2.1, it is immediate to conclude that the betting system above is fair if and only if $\mathrm{E}(G_{be}) = \mathrm{E}(G_{bo}) = 0 \Leftrightarrow p_i = o_i$, for any $i = 1, \ldots, k$. Conversely, if $o_i > p_i$, or

$p_i/o_i < 1$, meaning that the inverse odd set by the bookmaker for the event E_i is greater than the corresponding event probability, then we have a positive expected profit for the bookmaker, $\mathrm{E}(G_{bo}) > 0$, and a negative expected profit for the bettor, $\mathrm{E}(G_{be}) < 0$. Consider that in this artificial framework we are somehow assuming some true occurrence probabilities for the events of interest: however, in the majority of the applications, as in football, these are not accessible! Of course, the more precise they are estimated, and the more valuable will be the probabilistic evaluation.

Also from this market scenario we can conclude that the bookmaker can arbitrarily set and manage the inverse odds of a collection of mutually exclusive events in order to take favour, and then make money, against the bettor(s).

7.2.3 Market with more bookmakers and more events

Assume now that B bookmaker companies hold k mutually exclusive events $E_1, E_2, \dots .E_k$, and denote with $o_1^b, o_2^b, \dots, o_k^b$ the inverse odds vector for the b-th bookmaker, with $b = 1, 2, \dots, B$. Suppose again for simplicity that the bettor pays the prices $S_1, S_2, \dots, S_k$ to win one among the amounts $W_1^b = S_1/o_1^b, W_2^b = S_2/o_2^b, \dots, W_k^b = S_k/o_k^b$ in case one of the mutual events takes place: as a matter of convenience, we then assume that the bettor invests the same amounts $S_1, S_2, \dots, S_k$ for each of the B bookmakers. We can then compute the bettor's profit as:

$$\begin{aligned} G_{be}(E_i) &= \sum_{b=1}^{B}(S_i/o_i^b - \sum_{i=1}^{k} S_i) \times p_i \\ G_{be}(\bar{E}_i) &= -B(1-p_i) \times \sum_{i=1}^{k} S_i, \end{aligned} \tag{7.11}$$

whereas the bettor's expected profit is given by:

$$\begin{aligned} \mathrm{E}(G_{be}) &= \sum_{b=1}^{B}\sum_{i=1}^{k}\left[(S_i/o_i^b - \sum_{i=1}^{k} S_i) \times p_i - (1-p_i) \times \sum_{i=1}^{k} S_i\right] \\ &= \sum_{b=1}^{B}\sum_{i=1}^{k}\left[S_i\frac{p_i}{o_i^b} - \sum_{i=1}^{k} S_i\right]. \end{aligned} \tag{7.12}$$

For simplicity we focus on the single bookmaker: analogously as in Section 7.2.2, the market is fair, meaning $\mathrm{E}(G_{be}) = 0$, if and only if $p_i = o_i^b$ for any $i = 1, \dots, k$. If $p_i/o_i^b < 1$, then the bettor will have a negative expected profit when competing against the b-th bookmaker—however, in this market situation some compensations between the bookmakers' odds could occur. The symmetrical position for the distinct bookmakers, not reported here, can be easily derived as in the previous sections. As a final comment, we point out that the odds and then the inverse odds vary across the bookmakers—this happens in many sports, such as football, tennis, etc.: this practically means that a bettor could diversify the distinct bet amounts by looking for some "arbitrage" windows. For more details about the concept of arbitrage in financial markets, we refer the reader for instance to Dybvig and Ross (1989).

7.2.4 Bookmaker's gain in football

As shown above, the bookmakers inflate their probabilities by loading a margin ensuring a positive expected return. As an illustrative football example, suppose we want to place a bet on the English Premier League match Arsenal vs Manchester United, by putting 3 euros on one among the events: Arsenal win, draw, United win. The betting odds for the three events provided by an imaginary bookmaker are respectively: 2.5, 3, 3.2, which means that our eventual cash-flow is equal to 7.5, 9, or 9.6. If we take the inverse of the odds, 1/2.5, 1/3, 1/3.2, we realize that their sum exceeds one, being 1.046: this means that the bookmaker margin is equal to 4.6%, and this loading always makes the bet unfair for us, the bettor, and favourable to the bookies. The expected profit for the bettor is in fact:

$$\begin{aligned}(7.5-3)\times 0.4-3\times 0.646 &= -\,0.135 \quad \text{bet on home win}\\ (9-3)\times 0.333-3\times 0.7125 &= -\,0.1395 \quad \text{bet on draw}\\ (9.6-3)\times 0.3125-3\times 0.733 &= -\,0.1365 \quad \text{bet on away win,}\end{aligned}$$

which means that the bettor who assume the inverse odds as being "true" probabilities incurs in a probable loss, whereas the bookmaker, symmetrically, earns a probable gain. Suppose now to transform the inverse odds into coherent probabilities through the basic normalization procedure proposed in Section 7.1, which gives the following probabilities' vector for the match: $\pi = (0.382, 0.319, 0.299)$ for the Arsenal win, the draw, and the United win, respectively. If we adopt these probabilities for evaluating the outcomes corresponding to the Arsenal win, the draw, and the United win, the expected profit for the bettor will be:

$$\begin{aligned}(7.5-3)\times 0.382-3\times 0.618 &= -\,0.135 \quad \text{bet on home win}\\ (9-3)\times 0.319-3\times 0.681 &= -\,0.129 \quad \text{bet on draw}\\ (9.6-3)\times 0.299-3\times 0.701 &= -\,0.13 \quad \text{bet on away win.}\end{aligned}$$

Thus, according to the betting implied probabilities above, the expected profit is still always negative for the bettor, and, conversely, positive for the bookmaker: in fact, to consider the bookmaker's perspective it is just sufficient to take the symmetrical position and simply change the sign of the bet above. However, it is again worth noting that in frameworks such as football no one know the true probabilities.

7.3 Strategies on betting in football

7.3.1 Dixon and Coles approach

Once the vector of betting probabilities $\boldsymbol{\pi}$ has been obtained with one among the methods described in Section 7.1—either with basic normalization, Shin procedure, or regression analysis—we should try to assess how and when it is convenient to bet on some events *using the model probabilities.* The rationale behind betting is that if our model reflects approximately well the real chances of occurrence of a given event, then these probabilities should be used to challenge the bookmakers and eventually beat them: we have already seen in the previous sections that whenever $o_i > p_i$ for some binary events, then the bookmaker obtains a positive expected profit. Regarding football and the three-way outcomes, Dixon and Coles (1997) suggested to fix a *margin tolerance* δ, such that one would bet on a match/event i if and only if:

$$p_i^j/\pi_i^j > \delta, \quad j \in \Delta_i,$$

where p_i^j, π_i^j denote the probabilities for event i and occurrence j under the model and the bookmaker, respectively—we assume here to deal with a unique bookmaker, for simplicity Actually neither p_i^j nor π_i^j correspond to the true probabilities for match i, which are usually unknown in football and in general in sports; however, we could obtain a positive return if our estimated probabilities are sufficiently more accurate than those derived from the bookmakers, meaning their ratio exceeds a fixed tolerance δ. If the model probabilities p_i are accurate, then the expected gain from a unit bet for match i and outcome j is given by

$$\mathrm{E}(G_{be}) = p_i^j/\pi_i^j - 1. \tag{7.13}$$

As remarked by Dixon and Coles (1997), the choice of δ strongly depends on the amount of risk aversion undertaken by the bettor. They even propose to estimate δ or, alternatively, to monitor the return by varying the values of δ through a sensitivity analysis. If we increase δ, this means we adopt a stricter betting regime, but with fewer bets. Thus, the amount of δ strictly depends on the risk aversion of the single bettors: we give a practical example on the choice of δ in the case-study reported in Section 7.4.

7.3.2 Highest expected return

In a betting football strategy, two main questions arise: it is worth betting on a given match? If so, how much of the available resources should be invested for betting? In Section 7.1, we described two different procedures for inferring a vector of betting probabilities $\boldsymbol{\pi}$ from the inverse odds vector $\boldsymbol{o}$. The common expression "beating the bookmakers" may be interpreted in two distinct ways, from a probabilistic and a profitable point of view. According to the first definition, which is more appealing for statisticians, the π_i^j denote the betting probabilities provided by the bookmaker for the i-th game, with $j \in \Delta_i = \{\text{"}Home\ win\text{"}, \text{"}Draw\text{"}, \text{"}Away\ win\text{"}\}$. Additionally, let Y_{i1} and Y_{i2} denote the random variables representing the number of goals scored by two teams in the i-th match. If we adopt a Bayesian Poisson-based model among those introduced in Chapter 5, we can easily compute the following three-way model's posterior probabilities:

$$\begin{aligned}
p_i^{Home} &= P(Y_{i1} > Y_{i2}|\mathcal{D}),\\
p_i^{Draw} &= P(Y_{i1} = Y_{i2}|\mathcal{D}),\\
p_i^{Away} &= P(Y_{i1} < Y_{i2}|\mathcal{D}),
\end{aligned}$$

for each match i, conditioned on the past outcomes $\mathcal{D}$, by using the samples from the posterior predictive distribution. To assess whether we can probabilistically beat the bookmakers, we definitely need some probabilistic measures of performance, as those introduced in Chapter 3, such as the Brier score or the pseudo-R^2.

According to the second definition, "beating the bookmaker" means earning money by betting according to our model probabilities. One could bet one unit on the three-way match outcome with the highest expected return (Egidi et al., 2018b). Denoted with j_i and S_i the three-way outcome with the highest expected gain (7.13) and the money invested in the i-th match, respectively, the expected bettor profit is then defined, similarly as in (7.7), as:

$$\mathrm{E}(G_{be,i}) = S_i\left(\frac{p_i^{j_i}}{o_i^{j_i}} - 1\right), \tag{7.14}$$

where we again suppressed the bookmakers' index to ease the notation: it is evident how the expected profit is positive as long as $p_i^{j_i}/o_i^{j_i} > 1$, as remarked in Sections 7.2.1–7.2.3. Then, we extended a binary event scenario to a three-way event scenario.

7.3.3 Kelly approach

One of the fundamental steps for devising a good betting strategy is to establish the portion of available funds that the bettor is willing to invest on a sequence of events. The Kelly approach (Kelly, 1956) is a probabilistic strategy that determines the optimal amount of money for a single bet, and is given by

$$f^* = p - (1-p)/b, \tag{7.15}$$

where f^* is the fraction of bankroll to wager for the current bet, p is the probability of a win, and $b = 1/o - 1$ is the proportion of the bet gained with a win. The Kelly criterion can be extended to settings where the event outcomes are not restricted to be binary but, more generally, categorical, as in a football match. Unlikely for what happens in some gambling settings, where the probability of winning the bet can be known *a priori*, in a football match this is in general not accessible, and the best a bettor could do is to bet play by using some plausible subjective/model probabilities.

To illustrate the strategy with a binary example, suppose a bettor wants to invest some money on a horse race, by indicating the winner horse. Suppose he conjectures the winning probability for the selected horse is 0.2, whereas the odd for the horse provided by the bookmaker is 1:6.5, meaning the inverse odd is equal to 0.153, and $b = 1/o - 1 = 6.5 - 1 = 5.5$. Then, by applying the Equation (7.15), the ideal bankroll to wager is given by $f^* = 0.2 - (0.8)/5.5 = 0.055$, meaning that the bettor should bet 5.5% of the bankroll in favour of the selected horse. As a matter of gambling interpretation, if b equals $(1-p)/p$, then $f^* = 0$, then the criterion recommends the bettor to not invest any money on the given bet: in the horse race example, this would mean that $b = 0.8/0.2 = 4$, which corresponds to an betting odd of 1:5, less favourable than 1:6.5. If $b < (1-p)/p$ instead, then $f^* < 0$, meaning that the gambler should take the other side of the bet.

Consider another famous example, the French roulette, where the probabilities are known. Which is the ideal bankroll to wager for a red outcome? There are 37 numbers on the wheel, from 0 to 36, among which 18 are red numbers and 19 non-red numbers—the 0 is neither red nor black. Assume for simplicity that $b = 1$. The probability to win is known and given by $p = 18/37$, whereas $1-p = 19/37$. We compute the Kelly bankroll as $f^* = 18/37 - 19/37 = -1/37$, meaning the bettor should bet 1/37 of his/her bankroll that red will not come up. However, there is no explicit anti-red bet offered with comparable odds in roulette, so the best a Kelly gambler can do is bet nothing on French roulette!

We can easily extend the Kelly criterion to the football framework, where the main outcomes are: home win, draw, and away win. Consider again the example provided in Section 7.2.4, Arsenal vs Manchester United, where the inverse odds were: 1/2.5, 1/3, 1/3.2, then $b = 1.5, 2, 2.2$ in the three cases. If we were some optimistic Arsenal fans and we had estimated an Arsenal win probability of 0.45, the bankroll for the Arsenal win bet would be $f^* = 0.45 - 0.55/1.5 = 0.08$, meaning that we should invest 8% of our bankroll for the Arsenal win. What if we wanted to play with the probabilities derived from the bookmakers through, say, the basic normalization approach, 0.382, 0.319 and 0.299, respectively? Let's evaluate the bankrolls for the three cases:

$$\begin{aligned} f^* &= 0.382 - 0.618/1.5 = -0.03 \quad \text{bet on home win} \\ f^* &= 0.319 - 0.681/2 = \quad -0.02 \quad \text{bet on draw} \\ f^* &= 0.299 - 0.701/2.2 = -0.02 \quad \text{bet on away win,} \end{aligned}$$

all the three bankrolls would be negative, then we are suggested to not bet anything on this match: we got then another empirical confirmation of the favourite-longshot bias for the bookmakers, as explained in Section 7.2.4.

To summarise, the Kelly criterion is a useful approach to decide how much of the individual bankroll to wager on a given bet; nevertheless, the individual probabilities should be very accurate to guarantee to not lose money.

7.3.4 Expected profit optimization

According to a common sense the choice of the matches for which placing one or more bets depends on our utility for betting: we could decide which matches to bet on and obtain a favourable game to play as the posterior expected profit is positive. Epstein (2012) proposed to bet on outcomes with a positive expected profit but place the bets so we obtain a low variance of the profit. According to a similar perspective, Rue and Salvesen (2000) proposed to bet in order to maximize the expected profit while keeping the variance of the profit lower than some threshold.

An equivalent formulation is to maximise the expected profit minus the variance of the profit, which determine how we should place our bets up to a multiplicative constant. This constant can be found if we choose a specific value or an upper limit for the variance of the profit.

Let $\mathrm{E}(G_i^j)$ and $(\sigma_i^j)^2$ be the expected profit and the variance for betting an unit amount on outcome j in match i, where $j \in \Delta_i = \{$"*Home win*", "*Draw*", "*Away win*"$\}$, respectively: we can detect these values from the probabilities p_i^j and inverse odds o_i^j, as previously explained. For simplicity, suppose to not place more than one bet for each match, and let β_i^j be the corresponding bet. Let $U(\cdot)$ denote a proper bettor utility function, then the optimal bet is given by setting the condition:

$$\begin{aligned} &\underset{\beta_i^j \geq 0}{\text{argmax}}\, U(\{\beta_i^j\}), \quad \text{where} \\ U(\{\beta_i^j\}) = \mathrm{E}(\text{profit}) - \mathrm{Var}(\text{profit}) &= \sum_{i \in \beta} \beta_i^j (\mathrm{E}(G_i^j) - \beta_i^j (\sigma_i^j)^2). \end{aligned} \tag{7.16}$$

The analytical solution is given by $\beta_i^j = \max\{0, \mathrm{E}(G_i^j)/(2(\sigma_i^j)^2)\}$, where additionally we choose the outcome j with maximal $\beta_i^j \mathrm{E}(G_i^j)$ for match i in order to not place more than one bet for each match.

As an imaginary example, consider again the odds from the example considered in the previous sections, Arsenal vs Manchester United considered in the previous sections. For this match the expected profit for the single outcomes can be computed as $p^j/\pi^j - 1$, where p^j and π^j are the bettor and the betting probabilities for the outcome j, respectively. Suppose the bettor model probabilities for the home win, the draw and away win are: 0.41, 0.28, 0.31, whereas the bookmaker probabilities obtained through basic normalization are: 0.382, 0.319, 0.299. We then get the following expected profits from the bettor's perspective:

$$\begin{aligned} \mathrm{E}(G_{be}^H) &= 0.41/0.382 - 1 = 0.073 \\ \mathrm{E}(G_{be}^D) &= 0.28/0.319 - 1 = -0.122 \\ \mathrm{E}(G_{be}^L) &= 0.31/0.299 - 1 = 0.037. \end{aligned}$$

Under the case that $(\sigma_i^j)^2 = 0.5 \ \forall j$, Equation (7.12) implies the following optimal bets selection:

$$\begin{aligned}\beta^H &= \max\{0, 0.073/[2 \times (0.5)^2]\} &= 0.146\\ \beta^D &= \max\{0, -0.122/[2 \times (0.5)^2]\} &= 0\\ \beta^L &= \max\{0, 0.037/[2 \times (0.5)^2]\} &= 0.074.\end{aligned}$$

Additionally, we choose to bet on Arsenal win, since $\beta^H \mathrm{E}(G_{be}^H)$ is the highest value among the three possible combinations: then, the method suggests to place 14.6 cents on the Arsenal win.

7.4 Case Study: Italian Serie A 2009–2010

In this section we apply the predictive performance introduced in Chapter 3 on a real case study based on the Italian Serie A 2009/2010, the same season used in Chapters 4 and 5. Furthermore, the aim is to develop a sound betting strategy based on the probabilistic predictions from our models and the bookmakers' odds as collected by the website `https://www.football-data.co.uk/`.

As a comparison benchmark, we consider ten bookmakers: Bet365, Bet&Win, Blue Square, Gamebookers, Interwetten, Ladbrokes, Sportingbet, Stan James, VC Bet, and William Hill. The odds considered are referred as the *pre-closing* odds, in such a way that the odds for weekend games are collected Friday afternoons, and on Tuesday afternoons for midweek games: for each game we consider the home win, draw, and away win betting odds.

We adopt the following notation already introduced in Section 3.3 in Chapter 3: we denote the training set as $\mathcal{T} = \{1, 2, \dots, n^*\}$ and we will generate results for the remaining games, that is the test/held-out set $\mathcal{V} = \{n^*+1, n^*+2, \dots, n\}$. We outline two different scenarios: "mid-season predictions" with $n^* = 190$, and "leave-four-out" predictions, with $n^* = 340$, with the last four match-days to be forecasted. Predictive and betting performances—accuracy, Ranked Probability Score (RPS), Brier Score (BS), pseudo-R^2, and Average of Correct Probabilities (ACP)—see Section 3.4 in Chapter 3—are investigated for three considered Bayesian models, namely the double Poisson (DP, Section 4.4.1), the bivariate Poisson (BP, Section 4.4.2), and the diagonal-inflated bivariate Poisson—DIBP, Section 5.1.1 in Chapter 5—and for the ten aforementioned bookmakers. As clearly emerges from Tables 7.2 and 7.3, the predictive performances from the bookmakers—for which the betting odds are normalized according to Equation (7.1) in Section 7.1.1—are greater than those obtained under the three statistical models: as broadly remarked for instance by Štrumbelj (2014), it is empirically known that the betting odds represent the most accurate probabilistic evaluation for a football match. By comparing the two scenarios, it is also evident how the predictive performance, both for the bookmakers and statistical models, much improve as the training sample size increases, from $n^* = 190$ to $n^* = 340$: this confirms how both the actors, the model and the betting companies, *learn* along the season and formulate better probabilistic forecasts as the season evolves. The predictive performance in these tables have been computed through the `compare_foot()` function in the `footBayes` package. A minimal and illustrative code for the use of the function is provided here below in Code Snippet 46. The function output is reported in Output 43.

Code Snippet 46 Italian Serie A 2009/2010: predictive performances.

```
library(footBayes)
library(dplyr)

data(italy)  # available in the package
italy_2009 <- subset(italy[, c(2,3,4,6,7)], Season =="2009")
colnames(italy_2009) <- c("periods", "home_team", "away_team",
                          "home_goals", "away_goals") # rename!
n_train <- 190
n_test <- 190 # mid-season predictions

dp_stan <- stan_foot(data = italy_2009,
                     model="double_pois",
                     predict = n_test) # dp
bp_stan <- stan_foot(data = italy_2009,
                     model="biv_pois",
                     predict = n_test) # bp
bp_stan_infl <- stan_foot(data = italy_2009,
                          model="diag_infl_biv_pois",
                          predict = n_test) # inflated bp

test_data <- italy_2009[(n_train+1):(n_train+n_test), ]
compare_foot(list(dp = dp_stan, bp = bp_stan,
                  bp_infl = bp_stan_infl), test_data = test_data)
```

```
    Model     RPS   accuracy brier  pseudoR2  ACP
1 dp       0.2077   0.4789 0.6302   0.3513 0.3723
2 bp       0.2099   0.4789 0.6354   0.3480 0.3634
3 bp_infl  0.2083   0.4737 0.6317   0.3503 0.3695
```

Output 43: Italian Serie A 2009/2010: probabilistic predictive performance of three models.

In terms of betting actions, we use the probabilistic forecasts obtained from the double Poisson model—the model that yields the best predictive

TABLE 7.2
Italian Serie A 2009/2010, mid-season predictions ($n^* = 190$): measures of predictive performance. From the second column until the sixth one: Ranked Probability Score (RPS), accuracy (acc.), Brier Score (BS), pseudo-R^2, and Average of Correct Probabilities (ACP). Three considered models: double Poisson (DP), bivariate Poisson (BP), and diagonal-inflated bivariate Poisson (DIBP)

Model/bookie	RPS	acc.	BS	ps.-R^2	ACP
Bet365	0.19	0.54	0.58	0.38	0.41
Bet&Win	0.19	0.53	0.58	0.38	0.41
Blue Square	0.19	0.54	0.58	0.38	0.41
Gamebookers	0.19	0.54	0.58	0.38	0.41
Interwetten	0.19	0.54	0.58	0.38	0.41
Ladbrokes	0.19	0.54	0.58	0.38	0.41
Sportingbet	0.19	0.53	0.58	0.38	0.41
Stan James	0.19	0.54	0.58	0.38	0.41
VC Bet	0.19	0.54	0.58	0.38	0.41
William Hill	0.18	0.54	0.58	0.38	0.41
DP	0.21	0.48	0.63	0.35	0.37
BP	0.21	0.47	0.64	0.35	0.36
DIBP	0.21	0.47	0.63	0.35	0.37

performances in Output 43—and start with the Dixon & Coles strategy proposed in Section 7.3.1, by allowing for different values of the margin tolerance δ. As depicted in Figure 7.2 in the top panel plots, the joint choice of n^* and the margin tolerance has a relevant effect on the net profit: in fact, in the top left plot, with $n^* = 190$, a positive profit is obtained only for one bookmaker (William Hill) and only when the threshold δ in Section 7.3.1 is about 1.6. Instead, when the training sample size is $n^* = 340$, in the top right plot, a positive profit can be obtained from the majority of the bookmakers when δ is bounded between the values 1.2 and 1.6: VC Bet is the most "advantageous" bookmaker, since the profit we can obtain playing with its odds approximates the 20% when $\delta \approx 1.6$.

Another betting strategy could be adopted by following the highest expected return (HER) approach proposed in Section 7.3.2. Mid plots in Figure 7.2 depict the profit trends for both the training scenarios in function of the margin tolerance δ—note that in this approach, as remarked by Equation (7.14), the margin tolerance discriminates the ratios $p/o > \delta$, where as usual p denotes the model probability and o the inverse betting odd. Also according to this approach, the profit can be obtained quite easily in the second scenario (mid right plot), where VC Bet is again the most profitable betting company from a bettor's perspective, and is associated with a profit approximating 20% when $\delta \approx 1.45$.

Finally, the Kelly criterion in Section 7.3.3 is explored. As documented by many prominent experts, the Kelly criterion is very "aggressive" and could

TABLE 7.3
Italian Serie A 2009/2010, "leave-four-out predictions" ($n^* = 340$): measures of predictive performance. From the second column until the sixth one: Ranked Probability Score (RPS), accuracy (acc.), Brier Score (BS), pseudo-R^2, and Average of Correct Probabilities (ACP). Three considered models: double Poisson (DP), bivariate Poisson (BP), and diagonal-inflated bivariate Poisson (DIBP)

Model/Bookie	RPS	acc.	BS	ps.-R^2	ACP
Bet365	0.17	0.64	0.48	0.43	0.49
Bet&Win	0.17	0.64	0.49	0.43	0.49
Blue Square	0.17	0.64	0.48	0.43	0.49
Gamebookers	0.17	0.63	0.49	0.43	0.49
Interwetten	0.17	0.66	0.48	0.44	0.49
Ladbrokes	0.17	0.64	0.49	0.43	0.49
Sportingbet	0.17	0.65	0.49	0.43	0.49
Stan James	0.17	0.63	0.49	0.43	0.49
VC Bet	0.17	0.64	0.49	0.43	0.49
William Hill	0.17	0.64	0.48	0.43	0.49
DP	0.21	0.55	0.60	0.37	0.39
BP	0.21	0.57	0.60	0.37	0.38
DIBP	0.21	0.52	0.59	0.37	0.39

yield problematic trends in terms of final profits. In fact, the profit trends are reported against the margin tolerance γ—where we choose to bet on the outcomes such that $b \times p/(1-p) > \gamma$—in Figure 7.2, bottom plots. From this experiment, it is pretty impossible to yield a positive profit, and the trend decreases as the tolerance increases. In partial agreement with the other two approaches, we may note again that the profits are higher when $n^* = 340$—at least, the losses are less serious—in the bottom right plot, since the probabilistic forecasts are more accurate. We need also to note that deriving an optimal betting strategy for the Kelly criterion when dealing with multiple exclusive outcomes as in football is not an easy task, and many alternative algorithmic choices could be adopted. For this reason, we maintain that this third approach is just merely meant as an illustrative example, but it should not be considered as a real manner to make money from the bookmakers.

The final comment is that even if the probabilistic performance of our statistical models are not greater than those provided by the bookmakers operating in the market in terms of predictive performance, it is still possible to set up a profitable strategy to make money from them, as thoroughly shown through the Dixon & Coles and the highest expected return strategies.

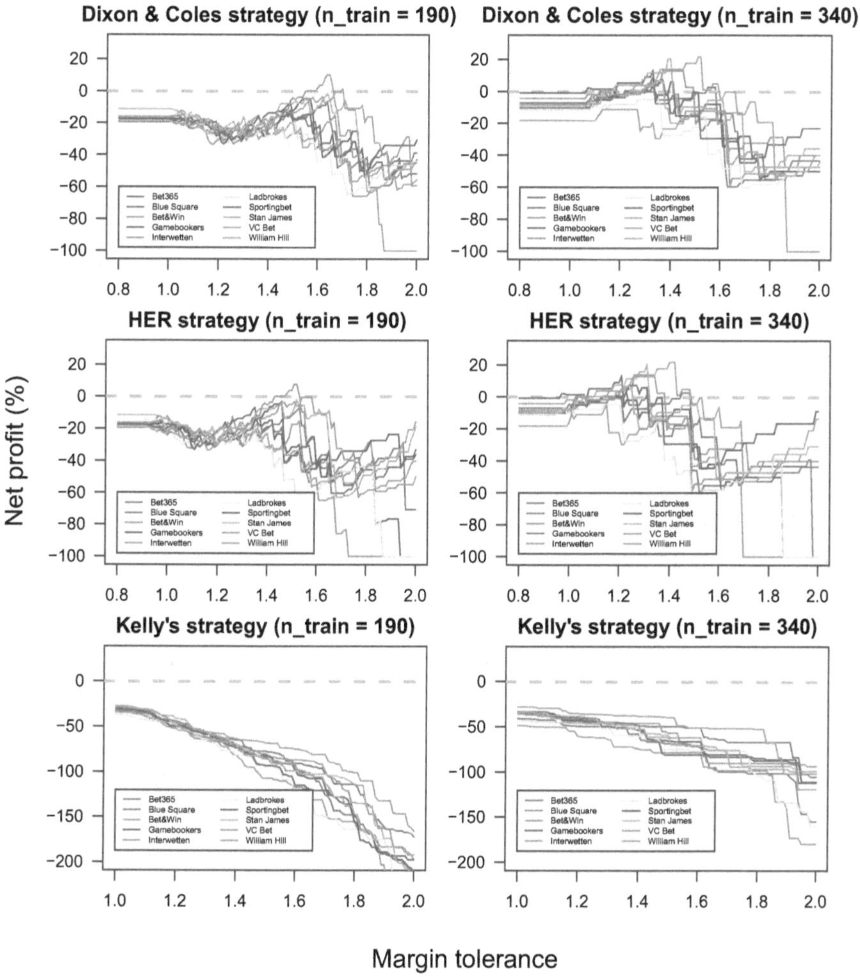

FIGURE 7.2
Italian Serie A 2009/2010, betting strategy using the double Poisson probabilistic forecasts and based on: the Dixon & Coles rule detailed in Section 7.3.1 (top plots); the highest expected return (HER) rule in 7.3.2 (mid plots); and the Kelly's approach in 7.3.3 (bottom plots). Two scenarios: mid-season predictions with $n^* = 190$ (left panels) and "leave-four-out" predictions with $n^* = 340$ (right panels). The single unit bet is 1 Dollar/Euro. The level of margin tolerance is plotted on the x-axis whereas the percentage profit is on the y-axis.

7.5 Summary and closing remarks of Chapter 7

Beating and making money from the bookmakers is usually seen as the most appealing task when producing football predictions, however this is often unrealistic. To challenge the bookmakers' predictions, one should understand how the usual odds relate to the probabilities of a given event, in this case a football match. In this chapter we provided an overview about the bookmakers' markets and some existing betting strategies as far as reviewing the current scientific literature; moreover, we provided a real case-study by using some predictive measures to assess the effectiveness of some model-based probabilistic predictions.

Section 7.1 explains how to transform the betting odds into probabilities by introducing the basic normalization in Section 7.1.1, the Shin's procedure in Section 7.1.2, and some regression analysis techniques in Section 7.1.3. The bookmaker market is then mathematically introduced in Section 7.2, with examples involving: one bookmaker and one event in Section 7.2.1, one bookmaker and more events in Section 7.2.2, and more bookmakers and more events in Section 7.2.3. The purpose of these section is to explain how to properly compute the expected profits, from both the bettor and the bookmaker perspective, and introduce the concept of fairness. Section 7.2.4 introduces the concept of gain, or margin, contained in the odds released by the bookies.

Some betting strategies are then provided in Section 7.3, such as the Dixon & Coles criterion in Section 7.3.1, the Kelly criterion introduced in Section 7.3.3, and the expected profit optimization in Section 7.3.4. Section 7.4 illustrates through a real case study focused on the Italian Serie A 2009/2010 the computation of various predictive performances arising in a football match by using the `footBayes` package and proposes some possible strategies to make money from the bookmakers when splitting a season in training and test sets.

Bibliography

Agresti, A. (2013). *Categorical Data Analysis.* Wiley and Sons, Hoboken, NJ, 3rd edition.

Agresti, A. (2015). *Foundations of Linear and Generalized Linear Models.* Wiley and Sons, Hoboken, NJ.

Akaike, H. (1973). "Information theory and an extension of the maximum likelihood principle". in B. Petrov and F. Csaki, eds., *Proceedings of 2nd International Symposium on Information Theory*, pages 267–281. Academiai Kiado, Budapest.

Albert, J. (2009). *Bayesian Computation with R.* Springer, Dordrecht.

Angelini, G. and De Angelis, L. (2017). Parx model for football match predictions. *Journal of Forecasting*, 36(7):795–807.

Anzer, G., Bauer, P., and Brefeld, U. (2021). The origins of goals in the German Bundesliga. *Journal of Sports Sciences*, 39(22):2525–2544.

Asif, M. and McHale, I. G. (2016). In-play forecasting of win probability in one-day international cricket: A dynamic logistic regression model. *International Journal of Forecasting*, 32(1):34–43.

Baboota, R. and Kaur, H. (2019a). Predictive analysis and modelling football results using machine learning approach for English premier league. *International Journal of Forecasting*, 35(2):741–755.

Baboota, R. and Kaur, H. (2019b). Predictive analysis and modelling football results using machine learning approach for English premier league. *International Journal of Forecasting*, 35(2):741–755.

Baio, G. and Blangiardo, M. (2010). Bayesian hierarchical model for the prediction of football results. *Journal of Applied Statistics*, 37(2):253–264.

Baker, R. and Scarf, P. (2020). Modifying Bradley–Terry and other ranking models to allow ties. *IMA Journal of Management Mathematics*, 32(4):451–463.

Barbieri, M. and Berger, J. (2004). 'Optimal predictive model selection'. *Annals of Statistics*, 32:870–897.

Barbiero, A. (2020). Discrete Weibull regression for modelling football outcomes. *International Journal of Business Intelligence and Data Mining*, 17(1):76–100.

Barnett, V. and Hilditch, S. (1993). The effect of an artificial pitch surface on home team performance in football (soccer). *Journal of the Royal Statistical Society Series A: Statistics in Society*, 156(1):39–50.

Bartlett, M. (1957). 'Comment on D.V. Lindley's statistical paradox'. *Biometrika*, 44:533–534.

Baxter, M. and Stevenson, R. (1988). Discriminating between the Poisson and negative binomial distributions: An application to goal scoring in association football. *Journal of Applied Statistics*, 15(3):347–354.

Benz, L. S. and Lopez, M. J. (2023). Estimating the change in soccer's home advantage during the covid-19 pandemic using bivariate Poisson regression. *AStA Advances in Statistical Analysis*, 107(1-2):205–232.

Bernardo, J. and Smith, A. (2000). *Bayesian Theory, 2nd edition.* Wiley, Chichester, UK.

Betancourt, M. (2017). A conceptual introduction to Hamiltonian Monte Carlo. *arXiv preprint arXiv:1701.02434*.

Betancourt, M. and Girolami, M. (2015a). Hamiltonian Monte Carlo for hierarchical models. *Current Trends in Bayesian Methodology with Applications*, 79(30):2–4.

Betancourt, M. and Girolami, M. (2015b). "Hamiltonian Monte Carlo for hierarchical models". in *Current Trends in Bayesian Methodology with Applications* (S. K. Upadhyay, U. Singh, D.K. Dey, and A. Loganathan, eds), pages (Ch. 4) 79–102. Chapman and Hall/CRC, New York.

Böhning, D., Dietz, E., Schlattmann, P., Mendonca, L., and Kirchner, U. (1999). The zero-inflated Poisson model and the decayed, missing and filled teeth index in dental epidemiology. *Journal of the Royal Statistical Society: Series A (Statistics in Society)*, 162(2):195–209.

Boshnakov, G., Kharrat, T., and McHale, I. G. (2017). A bivariate Weibull count model for forecasting association football scores. *International Journal of Forecasting*, 33(2):458–466.

Bradley, R. A. and Terry, M. E. (1952). Rank analysis of incomplete block designs: I. the method of paired comparisons. *Biometrika*, 39(3/4):324–345.

Brier, G. W. et al. (1950). Verification of forecasts expressed in terms of probability. *Monthly Weather Review*, 78(1):1–3.

Brillinger, D. R. (2008). Modelling game outcomes of the Brazilian 2006 series a championship as ordinal-valued. *Brazilian Journal of Probability and Statistics*, pages 89–104.

Brillinger, D. R. (2009). An analysis of Chinese super league partial results. *Science in China Series A: Mathematics*, 52(6):1139–1151.

Brooks, S. P. and Roberts, G. O. (1998). 'Assessing convergence of Markov chain Monte Carlo algorithms'. *Statistics and Computing*, 8:319–335.

Buraimo, B., Forrest, D., and Simmons, R. (2010). The 12th man?: refereeing bias in English and German soccer. *Journal of the Royal Statistical Society Series A: Statistics in Society*, 173(2):431–449.

Cain, M., Law, D., and Peel, D. (2002). Is one price enough to value a state-contingent asset correctly? Evidence from a gambling market. *Applied Financial Economics*, 12(1):33–38.

Cain, M., Law, D., and Peel, D. (2003). The favourite-longshot bias, bookmaker margins and insider trading in a variety of betting markets. *Bulletin of Economic Research*, 55(3):263–273.

Carlos, L.-P., Ezequiel, R., and Anton, K. (2019). How does video assistant referee (var) modify the game in elite soccer? *International Journal of Performance Analysis in Sport*, 19(4):646–653.

Carpenter, B., Gelman, A., Hoffman, M., Lee, D., Goodrich, B., Betancourt, M., Brubaker, M., Guo, J., Li, P., and Riddell, A. (2017). Stan: A Probabilistic Programming Language. *Journal of Statistical Software*, 76.

Carpita, M., Ciavolino, E., and Pasca, P. (2019). Exploring and modelling team performances of the Kaggle European soccer database. *Statistical Modelling*, 19(1):74–101.

Carpita, M., Sandri, M., Simonetto, A., and Zuccolotto, P. (2015). Discovering the drivers of football match outcomes with data mining. *Quality Technology & Quantitative Management*, 12(4):561–577.

Casas, A. and Fawaz, Y. (2016). Altitude as handicap in rank-order football tournaments. *Applied Economics Letters*, 23(3):180–183.

Cattelan, M., Varin, C., and Firth, D. (2013). Dynamic Bradley–Terry modelling of sports tournaments. *Journal of the Royal Statistical Society Series C: Applied Statistics*, 62(1):135–150.

Chumacero, R. A. (2009). Altitude or hot air? *Journal of Sports Economics*, 10(6):619–638.

Clarke, S. R. and Norman, J. M. (1995). Home ground advantage of individual clubs in English soccer. *Journal of the Royal Statistical Society: Series D (The Statistician)*, 44(4):509–521.

Cohen, J. (1960). A coefficient of agreement for nominal scales. *Educ and Psychological Measurement*, 20(1):37–46.

Constantinou, A. C. and Fenton, N. E. (2012a). Solving the problem adequate scoring rules for assessing probabilistic football forecast n *Journal of Quantitative Analysis in Sports*, 8(1).

Constantinou, A. C. and Fenton, N. E. (2012b). Solving the problem adequate scoring rules for assessing probabilistic football forecast r *Journal of Quantitative Analysis in Sports*, 8(1).

Cowles, M. and Carlin, B. (1996). 'Markov chain Monte Carlo conve diagnostics: A comparative review'. *Journal of the American Sta Association*, 91:883–904.

Crowder, M., Dixon, M., Ledford, A., and Robinson, M. (2002). Dynami elling and prediction of english football league matches for betting. *of the Royal Statistical Society Series D: The Statistician*, 51(2):157

Davidson, R. R. and Beaver, R. J. (1977). On extending the Bradley model to incorporate within-pair order effects. *Biometrics*, pages 69

Davidson, R. R. J. (1970). On extending the Bradley-Terry model to modate ties in paired comparison experiments. *Journal of the Am Statistical Association*, 65:317–328.

Dawson, P., Dobson, S., Goddard, J., and Wilson, J. (2007). Are footba ees really biased and inconsistent?: evidence on the incidence of disci sanction in the english premier league. *Journal of the Royal Statist ciety Series A: Statistics in Society*, 170(1):231–250.

de Finetti, B. (1931). Sul significato soggettivo della probabilità. *Fund Mathematicae*, 17:298–329.

De Finetti, B. (1970). Logical foundations and measurement of sub probability. *Acta Psychological*, 34:129–145.

Deb, S. (2022). A goal based index to analyze the competitive balar football league. *Journal of Quantitative Analysis in Sports*, 18(3):1

Dempster, A., Laird, N., and Rubin, D.B. (1977). Maximum Likelihoo Incomplete Data via the EM Algorithm. *Journal of the Royal St Society. Series B (Methodological)*, 39:1–38.

Diniz, M. A., Izbicki, R., Lopes, D., and Salasar, L. E. (2019). Con probabilistic predictive models applied to football. *Journal of the tional Research Society*, 70(5):770–782.

Dixon, M. and Coles, S. (1997). Modelling association football sco inefficiencies in football betting market. *Journal of the Royal St Society C*, 46:265–280.

Dixon, M. and Robinson, M. (1998). A birth process model for association football matches. *Journal of the Royal Statistical Society: Series D (The Statistician)*, 47(3):523–538.

Dobson, S., Goddard, J., et al. (2017). Evaluating probabilities for a football in-play betting market. *The Economics of Sports Betting*, pages 52–70.

Dobson, S., Goddard, J. A., and Dobson, S. (2001). *The economics of football*, volume 10. Cambridge University Press, Cambridge.

Dowie, J. (1982). Why Spain should win the world cup. *New Scientist*, 94(10):693–695.

Dybvig, P. H. and Ross, S. A. (1989). Arbitrage. In *Finance*, pages 57–71. Springer.

Egidi, L. and Gabry, J. (2018). Bayesian hierarchical models for predicting individual performance in soccer. *Journal of Quantitative Analysis in Sports*, 14(3):143–157.

Egidi, L., Pauli, F., and Torelli, N. (2018a). Are the shots predictive for the football results?

Egidi, L., Pauli, F., and Torelli, N. (2018b). Combining historical data and bookmakers' odds in modelling football scores. *Statistical Modelling*, 18(5-6):436–459.

Egidi, L. and Torelli, N. (2021a). Comparing goal-based and result-based approaches in modelling football outcomes. *Social Indicators Research*, 156:801–813.

Epstein, E. S. (1969). A scoring system for probability forecasts of ranked categories. *Journal of Applied Meteorology*, 8(6):985–987.

Epstein, R. A. (2012). *The Theory of Gambling and Statistical Logic*. Academic Press.

Erkanli, A. (1994). 'Laplace approximations for posterior expectation when the model occurs at the boundary of the parameter space'. *Journal of the American Statistical Association*, 89:205–258.

Evans, M. and Swartz, T. (1996). 'Discussion of methods for approximating integrals in statistics with special emphasis on Bayesian integration problems'. *Statistical Science*, 11:54–64.

Fagerland, M. W., Hosmer, D. W., and Bofin, A. M. (2008). Multinomial goodness-of-fit tests for logistic regression models. *Statistics in Medicine*, 27:4238–4253.

Fischer, K. and Haucap, J. (2021). Does crowd support drive the home advantage in professional football? evidence from German ghost games during the covid-19 pandemic. *Journal of Sports Economics*, 22(8):982–1008.

Gabry, J., Češnovar, R., Johnson, A., and Bronder, S. (2024). *cmdstanr: R Interface to 'CmdStan'.* R package version 0.8.1, https://discourse.mc-stan.org.

Gan, N. (2000). General zero-inflated models and their applications.

Gelfand, A., Hills, S., Racine-Poon, A., and Smith, A. (1990). 'Illustration of Bayesian inference in normal data models using Gibbs sampling'. *Journal of the American Statistical Association*, 85:972–985.

Gelfand, A. and Smith, A. (1990). 'Sampling-based approaches to calculating marginal densities'. *Journal of the American Statistical Association*, 85. 398–409.

Gelman, A. (2006). Prior distributions for variance parameters in hierarchical models (comment on article by Browne and Draper).

Gelman, A. (2014). Stan goes to the world cup.

Gelman, A., Carlin, J. B., Stern, H. S., Dunson, D. B., Vehtari, A., and Rubin, D. B. (2013). *Bayesian Data Analysis.* Chapman and Hall/CRC.

Gelman, A. and Hill, J. (2006). *Data Analysis using Regression and Multilevel/hierarchical Models.* Cambridge University Press.

Gelman, A., Hwang, J., and Vehtari, A. (2014). Understanding predictive information criteria for Bayesian models. *Statistics and Computing*, 24(6):997–1016.

Gelman, A., Jakulin, A., Pittau, M. G., and Su, Y.-S. (2008). A weakly informative default prior distribution for logistic and other regression models. *Annals of Applied Statistics*, 2(4):1360–1383.

Gelman, A., Meng, X.-L., and Stern, H. (1996). Posterior predictive assessment of model fitness via realized discrepancies. *Statistica Sinica*, pages 733–760.

Gelman, A., Rubin, D. B., et al. (1992). Inference from iterative simulation using multiple sequences. *Statistical Science*, 7(4):457–472.

Geman, S. and Geman, D. (1984a). 'Stochastic relaxation, Gibbs distributions and the Bayesian restoration of images'. *IEEE Transactions on Pattern Analysis and Machine Intelligence*, 6:721–741.

Geman, S. and Geman, D. (1984b). Stochastic relaxation, Gibbs distributions, and the Bayesian restoration of images. *IEEE Transactions on Pattern Analysis and Machine Intelligence*, (6):721–741.

Genest, C. and Nešlehová, J. (2007). A primer on copulas for count dat *ASTIN Bulletin: The Journal of the IAA*, 37(2):475–515.

Gilks, W., Richardson, S., and Spiegelhalter, D. (1996). *Markov Chain Mon Carlo in Practice*. Interdisciplinary Statistics. Chapman & Hall, Suffol UK.

Giulianotti, R. and Robertson, R. (2004). The globalization of football: study in the glocalization of the 'serious life'. *British Journal of Sociolog* 55:545–568.

Givens, G. H. and Hoeting, J. A. (2012). *Computational Statistics*. Joh Wiley & Sons, Hoboken, NJ, USA, 2nd edition.

Glenn, W. A. and David, H. A. (1960). Ties in paired-comparison experimen using a modified Thurstone–Mosteller model. *Biometrics*, 16:86–109.

Goddard, J. (2005a). Regression models for forecasting goals and mat results in association football. *International Journal of Forecastin* 21(2):331–340.

Goes, F., Meerhoff, L., Bueno, M., Rodrigues, D., Moura, F., Brink, M Elferink-Gemser, M., Knobbe, A., Cunha, S., Torres, R., et al. (2021). U locking the potential of big data to support tactical performance analys in professional soccer: A systematic review. *European Journal of Sport Sc ence*, 21(4):481–496.

Goes, K. (2021). Estimating the most important football player statisti using neural networks. B.S. thesis.

Groll, A. and Abedieh, J. (2013). Spain retains its title and sets a new recor generalized linear mixed models on European football championships. *Jou nal of Quantitative Analysis in Sports*, 9(1):51–66.

Groll, A., Hvattum, L. M., Ley, C., Popp, F., Schauberger, G., Van Eetveld H., and Zeileis, A. (2021). Hybrid machine learning forecasts for the UEF Euro 2020. *arXiv preprint arXiv:2106.05799*.

Groll, A., Kneib, T., Mayr, A., and Schauberger, G. (2016). Who's th favourite?–a bivariate Poisson model for the UEFA European football cham pionship 2016.

Groll, A., Kneib, T., Mayr, A., and Schauberger, G. (2018a). On the d pendency of soccer scores–a sparse bivariate Poisson model for the UEF European football championship 2016. *Journal of Quantitative Analysis Sports*, 14(2):65–79.

Groll, A., Ley, C., Schauberger, G., and Van Eetvelde, H. (2018b). Predictio of the fifa world cup 2018-a random forest approach with an emphasis o estimated team ability parameters. *arXiv preprint arXiv:1806.03208*.

Groll, A., Schauberger, G., and Tutz, G. (2015). Prediction of ma ternational soccer tournaments based on team-specific regularized F regression: An application to the FIFA world cup 2014. *Journal of Q tative Analysis in Sports*, 11(2):97–115.

Harvey, A. C. and Fernandes, C. (1989). Time series models for count o itative observations. *Journal of Business & Economic Statistics*, 7(4 417.

Hastings, W. (1970). 'Monte Carlo sampling methods using Markov and their applications'. *Biometrika*, 57:97–109.

Henery, R. J. (1992). An extension to the Thurstone-Mosteller mo chess. *Journal of the Royal Statistical Society Series D: The Stati* 41(5):559–567.

Heuer, A. and Rubner, O. (2009). Fitness, chance, and myths: an ob view on soccer results. *The European Physical Journal B*, 67:445–45

Hosmer, D. W., Hosmer, T., le Cessie, S., and Lemeshow, S. (1997). parison of goodness-of-fit tests for the logistic regression model. *St in Medicine*, 16:965–980.

Hosmer, D. and Lemeshow, S. (1980). A goodness-of-fit test for n logistic regression model. *Communications in Statistics A*, 9:1043–1

Hvattum, L. M. and Arntzen, H. (2010). Using elo ratings for ma sult prediction in association football. *International Journal of Fore* 26(3):460–470.

Johnson, N. L., Kotz, S., and Balakrishnan, N. (1997). *Discrete Mult Distributions*, volume 165. Wiley New York.

Jullien, B., Salanié, B., et al. (1994). Measuring the incidence of inside ing: a comment on Shin. *Economic Journal*, 104(427):1418–1419.

Karlis, D. and Ntzoufras, I. (2000a). On modelling soccer data. S 3:229–244.

Karlis, D. and Ntzoufras, I. (2000b). On modelling soccer data. S 3(4):229–244.

Karlis, D. and Ntzoufras, I. (2003). Analysis of sports data by using bi Poisson models. *Journal of the Royal Statistical Society D*, 52(3):38

Karlis, D. and Ntzoufras, I. (2006). Bayesian analysis of the differe count data. *Statistics in Medicine*, 25(11):1885–1905.

Karlis, D. and Ntzoufras, I. (2009). Bayesian modelling of football ou using the Skellam's distribution for the goal difference. *IMA Jou Management Mathematics*, 20:133–146.

Kass, R. E. and Raftery, A. E. (1995). Bayes factors. *Journal of the American Statistical Association*, 90(430):773–795.

Kass, R. E. and Wasserman, L. (1995). A reference Bayesian test for nested hypotheses and its relationship to the Schwarz criterion. *Journal of American Statistical Association*, 90:928–934.

Kelly, J. L. (1956). A new interpretation of information rate. *The Bell System Technical Journal*, 35(4):917–926.

Kharratzadeh, M. (2017). Hierarchical Bayesian modeling of the english premier league.

Klemp, M., Wunderlich, F., and Memmert, D. (2021). In-play forecasting in football using event and positional data. *Scientific Reports*, 11(1):24139.

Kocherlakota, S. and Kocherlakota, K. (2017). *Bivariate Discrete Distributions*. CRC Press.

Koopman, S. J. and Lit, R. (2015). A dynamic bivariate Poisson model for analysing and forecasting match results in the English Premier League. *Journal of the Royal Statistical Society A*, 178:167–186.

Koopman, S. J. and Lit, R. (2019a). Forecasting football match results in national league competitions using score-driven time series models. *International Journal of Forecasting*, 35(2):797–809.

Kuk, Y. C. (1995). Modelling paired comparison data with large numbers of draws and large variability of draw percentages among players. *Statistician*, 44:523–528.

Lago-Peñas, C., Gómez, M., and Pollard, R. (2021). The effect of the video assistant referee on referee's decisions in the Spanish Laliga. *International Journal of Sports Science & Coaching*, 16(3):824–829.

Lambert, D. (1992). Zero-inflated Poisson regression, with an application to defects in manufacturing. *Technometrics*, 34(1):1–14.

Landis, J. R. and Koch, G. G. (1977). The measurement of observer agreement for categorical data. *Biometrics*, 33(1):159–174.

Lee, A. (1997). Modeling scores in the Premier League: Is Manchester United really the best? *Chance*, 10:15–19.

Lee, A. (1999). Applications: Modelling rugby league data viabivariate negative binomial regression. *Australian & New Zealand Journal of Statistics*, 41(2):141–152.

Li, C.-S., Lu, J.-C., Park, J., Kim, K., Brinkley, P. A., and Peterson, J. P. (1999). Multivariate zero-inflated Poisson models and their applications. *Technometrics*, 41(1):29–38.

Lindley, D. (1957). 'A statistical paradox'. *Biometrika*, 44:187–192.

Lunn, D., Jackson, C., Best, N., Thomas, A., and Spiegelhalter, D. (2013). *The BUGS Book: A Practical Introduction to Bayesian Analysis*. Chapman and Hall/CRC, New York.

Lunn, D., Thomas, A., Best, N., and Spiegelhalter, D. (2000). WinBUGS — a Bayesian modelling framework: concepts, structure, and extensibility. *Statistics and Computing*, 10:325–337.

Macrì Demartino, R., Egidi, L., and Torelli, N. (2024). Alternative ranking measures to predict international football results. *Computational Statistics*, pages 1–19.

Maher, M. (1982). Modelling association football scores. *Statistica Neerlandica*, 36:109–118.

Manderson, A., Murray, K., and Turlach, B. (2018). Dynamic Bayesian forecasting of afl match results using the skellam distribution. *Australian & New Zealand Journal of Statistics*, 60(2):174–187.

Martin, A. D., Quinn, K. M., and Park, J. H. (2011). MCMCpack: Markov chain monte carlo in R. *Journal of Statistical Software*, 42(9):22.

Mattera, R. (2023). Forecasting binary outcomes in soccer. *Annals of Operations Research*, 325(1):115–134.

McHale, I. and Scarf, P. (2007). Modelling soccer matches using bivariate discrete distributions with general dependence structure. *Statistica Neerlandica*, 61(4):432–445.

McHale, I. and Scarf, P. (2011a). Modelling the dependence of goals scored by opposing teams in international soccer matches. *Statistical Modelling*, 11(3):219–236.

McHale, I. and Scarf, P. (2011b). Modelling the dependence of goals scored by opposing teams in international soccer matches. *Statistical Modelling*, 11(3):219–236.

McShane, B., Adrian, M., Bradlow, E. T., and Fader, P. S. (2008). Count models based on Weibull interarrival times. *Journal of Business & Economic Statistics*, 26(3):369–378.

Mechtel, M., Bäker, A., Brändle, T., and Vetter, K. (2011). Red cards: Not such bad news for penalized guest teams. *Journal of Sports Economics*, 12(6):621–646.

Metropolis, N., Rosenbluth, A., Rosenbluth, M., Teller, A., and Teller, E. (1953). 'Equations of state calculations by fast computing machine'. *Journal of Chemical Physics*, 21:1087–1092.

Moroney, M. J. (1956). *Facts from Figures*, volume 236. Penguin books Harmondsworth, Middlesex.

Murphy, A. H. (1970). The ranked probability score and the probability score: A comparison. *Monthly Weather Review*, 98(12):917–924.

Myers, B. R. (2012). A proposed decision rule for the timing of soccer substitutions. *Journal of Quantitative Analysis in Sports*, 8(1).

Nadarajah, S. and Chan, S. (2018). Discrete distributions based on inter arrival times with application to football data. *Communications in Statistics-Theory and Methods*, 47(1):147–165.

Neave, N. and Wolfson, S. (2003). Testosterone, territoriality, and the 'home advantage'. *Physiology & Behavior*, 78(2):269–275.

Nelder, J. A. and Wedderburn, R. W. (1972). Generalized linear models. *Journal of the Royal Statistical Society Series A: Statistics in Society*, 135(3):370–384.

Nelsen, R. B. (2006). *An Introduction to Copulas*. Springer.

Nevill, A. M., Balmer, N. J., and Williams, A. M. (2002). The influence of crowd noise and experience upon refereeing decisions in football. *Psychology of Sport and Exercise*, 3(4):261–272.

Nevill, A. M., Newell, S. M., and Gale, S. (1996). Factors associated with home advantage in english and Scottish soccer matches. *Journal of Sports Sciences*, 14(2):181–186.

Nevo, D. and Ritov, Y. (2013). Around the goal: examining the effect of the first goal on the second goal in soccer using survival analysis methods. *Journal of Quantitative Analysis in Sports*, 9(2):165–177.

Nikoloulopoulos, A. K. (2013). Copula-based models for multivariate discrete response data. In *Copulae in Mathematical and Quantitative Finance: Proceedings of the Workshop Held in Cracow, 10-11 July 2012*, pages 231–249. Springer.

Ntzoufras, I. (2009). *Bayesian Modeling Using WinBUGS*. Wiley Series in Computational Statistics. Hoboken, NJ.

Ntzoufras, I. (2011). *Bayesian Modeling using WinBUGS*, volume 698. John Wiley & Sons.

Ötting, M. and Karlis, D. (2023). Football tracking data: a copula-based hidden Markov model for classification of tactics in football. *Annals of Operations Research*, 325(1):167–183.

Owen, A. (2011). Dynamic Bayesian forecasting models of football outcomes with estimation of the evolution variance parameter. *IMA J of Management Mathematics*, 22:99–113.

Park, J. H., Cameletti, M., Pang, X., and Quinn, K. M. (2023). Cra view: Bayesian inference (version 2023-07-17). *CRAN Task View.*

Pawlowski, T. and Nalbantis, G. (2019). Competitive balance: Measu and relevance. *The SAGE Handbook of Sports Economics*, pages 15

Pearson, M., Livingston, G., Jr, and King, R. (2020). An explora predictive football modelling. *Journal of Quantitative Analysis in* 16(1):27–39.

Peng, R. D. (2022). *Advanced Statistical Computing.* Online available at `bookdown.org`, available at `https://bookdown.org/r` `advstatcomp/`.

Plummer, M. (2003). JAGS: A Program for Analysis of Bayesian ical Models Using Gibbs Sampling. *Proceedings of the 3rd Intern Workshop on Distributed Statistical Computing (DSC 2003), March Vienna, Austria. ISSN 1609-395X.*

Plummer, M., Best, N., Cowles, K., and Vines, K. (2006). 'CODA: C gence diagnosis and output analysis for MCMC'. *R News*, 6(1):7–11 able at `http://CRAN.R-project.org/doc/Rnews/Rnews_2006-1.p`

Pollard, R. (1986). Home advantage in soccer: A retrospective analysis *nal of Sports Sciences*, 4(3):237–248.

Rao, P. V. and Kupper, L. L. (1970). Ties in paired-comparison experi A generalization of the Bradley-Terry model. *Journal of the A Statistical Association*, 62:194–204.

Rathke, A. (2017). An examination of expected goals and shot effici soccer. *Journal of Human Sport and Exercise*, 12(2):514–529.

Reep, C. and Benjamin, B. (1968). Skill and chance in association f *Journal of the Royal Statistical Society. Series A (General)*, 131(4):5

Reep, C., Pollard, R., and Benjamin, B. (1971). Skill and chance games. *Journal of the Royal Statistical Society Series A: Statistics ety*, 134(4):623–629.

Ridder, G., Cramer, J. S., and Hopstaken, P. (1994). Down to ten: E ing the effect of a red card in soccer. *Journal of the American St Association*, 89(427):1124–1127.

Rissanen, J. (1986). Order estimation by accumulated prediction errors *nal of Applied Probability*, 23:55–61.

Robert, C. and Casella, G. (2013). *Monte Carlo Statistical Methods*. Springer Science & Business Media.

Rossi, A., Pappalardo, L., Cintia, P., Fernández, J., Iaia, M. F., Medina, D., et al. (2017). Who is going to get hurt? predicting injuries in professional soccer. In *MLSA@ PKDD/ECML*, pages 21–30.

Rue, H. and Salvesen, O. (2000). Prediction and retrospective analysis of soccer matches in a league. *Journal of the Royal Statistical Society D*, 49:399–418.

Scarf, P., Khare, A., and Alotaibi, N. (2022). On skill and chance in sport. *IMA Journal of Management Mathematics*, 33(1):53–73.

Scarf, P. and Rangel Jr, J. S. (2017). Models for outcomes of soccer matches. In *Handbook of Statistical Methods and Analyses in Sports*, pages 357–370. Chapman and Hall/CRC.

Schauberger, G., Groll, A., and Tutz, G. (2016). Modeling football results in the German Bundesliga using match-specific covariates. *Technical Report, Department of Statistics, University of Munich*, 197:77–95.

Schauberger, G., Groll, A., and Tutz, G. (2018a). Analysis of the importance of on-field covariates in the German Bundesliga. *Journal of Applied Statistics*, 45(9):1561–1578.

Schwarz, G. (1978). Estimating the dimension of a model. *Annals of Statistics*, 6:461–464.

Shahtahmassebi, G. and Moyeed, R. (2016). An application of the generalized Poisson difference distribution to the Bayesian modelling of football scores. *Statistica Neerlandica*, 70(3):260–273.

Shao, J. (1993). Linear model selection by cross-validation. *Journal of American Statistical Association*, 88:486–494.

Shao, J. (1997). An asymptotically theory for linear model selection (with discussion). *Statistica Sinica*, 7:221–264.

Shin, H. S. (1991). Optimal betting odds against insider traders. *The Economic Journal*, 101(408):1179–1185.

Shin, H. S. (1993). Measuring the incidence of insider trading in a market for state-contingent claims. *The Economic Journal*, 103(420):1141–1153.

Silva, R. M. and Swartz, T. B. (2016). Analysis of substitution times in soccer. *Journal of Quantitative Analysis in Sports*, 12(3):113–122.

Skellam, J. G. (1946). The frequency distribution of the difference between two Poisson variates belonging to different populations. *Journal of the Royal Statistical A*, 109:296.

Smit, R. C., Ravazzolo, F., Rossini, L., et al. (2020). Dynamic Bayesian forecasting of english premier league match results with the skellam distribution. Technical report, Faculty of Economics and Management at the Free University of Bozen.

Smith, B. J. (2007). boa: An r package for mcmc output convergence assessment and posterior inference. *Journal of Statistical Software*, 21(11):1–37.

Smith, M. A., Paton, D., and Williams, L. V. (2009). Do bookmakers possess superior skills to bettors in predicting outcomes? *Journal of Economic Behavior & Organization*, 71(2):539–549.

Sors, F., Grassi, M., Agostini, T., and Murgia, M. (2021). The sound of silence in association football: Home advantage and referee bias decrease in matches played without spectators. *European journal of sport science*, 21(12):1597–1605.

Spiegelhalter, D. and Ng, Y.-L. (2009). One match to go! *Significance*, 6(4):151–153.

Spiegelhalter, D., Thomas, A., Best, N., and Gilks, W. (1996). *BUGS 0.5: Bayesian Inference Using Gibbs Sampling Manual.* MRC Biostatistics Unit, Institute of Public Health, Cambridge, UK.

Spiegelhalter, D., Thomas, A., Best, N., and Lunn, D. (2003). *WinBUGS User Manual,* Version 1.4. MRC Biostatistics Unit, Institute of Public Health and Department of Epidemiology and Public Health, Imperial College School of Medicine, UK, available at `http://www.mrc-bsu.cam.ac.uk/bugs`.

Spiegelhalter, D. J., Best, N. G., Carlin, B. P., and Van Der Linde, A. (2002). Bayesian measures of model complexity and fit. *Journal of the Royal Statistical Society B*, 64(4):583–639.

Spitz, J., Wagemans, J., Memmert, D., Williams, A. M., and Helsen, W. F. (2021). Video assistant referees (var): The impact of technology on decision making in association football referees. *Journal of Sports Sciences*, 39(2):147–153.

Stan Development Team (2022). RStan: the R interface to Stan. R package version 2.21.7.

Stefani, R. T. (1980). Improved least squares football, basketball, and soccer predictions. *IEEE Transactions on Systems, Man, and Cybernetics*, 10(2):116–123.

Stefani, R. T. (1983). Observed betting tendencies and suggested betting strategies for European football pools. *Journal of the Royal Statistical Society: Series D (The Statistician)*, 32(3):319–329.

Stone, M. (1977). 'An asymptotic equivalence of choice of model by cross-validation and Akaike's criterion'. *Journal of the Royal Statistical Society B*, 39:44–47.

Štrumbelj, E. (2014). On determining probability forecasts from betting odds. *International Journal of Forecasting*, 30(4):934–943.

Szymanski, S. (2003). The economic design of sporting contests. *Journal of Economic Literature*, 41(4):1137–1187.

Thomas, A. C. (2007). Inter-arrival times of goals in ice hockey. *Journal of Quantitative Analysis in Sports*, 3(3).

Tibshirani, R. (1996). Regression shrinkage and selection via the lasso. *Journal of the Royal Statistical Society Series B: Statistical Methodology*, 58(1):267–288.

Tierney, L. and Kadane, J. (1986). 'Accurate approximations for posterior moments and marginal densities'. *Journal of the American Statistical Association*, 81:82–86.

Tierney, L., Kass, R., and Kadane, J. (1989). 'Fully exponential Laplace approximations to expectations and variances of nonpositive functions'. *Journal of the American Statistical Association*, 84:710–716.

Titman, A., Costain, D., Ridall, P., and Gregory, K. (2015). Joint modelling of goals and bookings in association football. *Journal of the Royal Statistical Society Series A: Statistics in Society*, 178(3):659–683.

Train, K. E. (2009). *Discrete Choice Methods with Simulation.* Cambridge university press.

Trombley, M. J. (2016). Does artificial grass affect the competitive balance in major league soccer? *Journal of Sports Analytics*, 2(2):73–87.

Tsokos, A., Narayanan, S., Kosmidis, I., Baio, G., Cucuringu, M., Whitaker, G., and Király, F. (2019a). Modeling outcomes of soccer matches. *Machine Learning*, 108:77–95.

Tzai, L., Ntzoufras, I., Vassalos, V., and Dimitriou, S. (2021). Predictions of European Basketball Match Results with Machine Learning Algorithms. *Research Preprint, Athens University of Economics and Business.*

Ulmer, B., Fernandez, M., and Peterson, M. (2013). Predicting soccer match results in the english premier league. *Doctoral dissertation, Doctoral dissertation, Ph.D. dissertation, Stanford.*

Van Eetvelde, H., Hvattum, L., and Ley, C. (2021). The Probabilistic Final Standing Calculator: a fair stochastic tool to handle abruptly stopped football seasons. *AStA Advances in Statistical Analysis*; *https://doi.org/10.1007/s10182-021-00416-6*.

Vehtari, A., Gelman, A., and Gabry, J. (2017). Practical Bayesian model evaluation using leave-one-out cross-validation and waic. *Statistics and Computing*, 27(5):1413–1432.

Vehtari, A., Simpson, D., Gelman, A., Yao, Y., and Gabry, J. (2015). Pareto smoothed importance sampling. *arXiv preprint arXiv:1507.02646*.

Volf, P. (2009). A random point process model for the score in sport matches. *IMA Journal of Management Mathematics*, 20(2):121–131.

Wald, A. (1943). Tests of statistical hypotheses concerning several parameters when the number of observations is large. *Transactions of the American Mathematical Society*, 54:426–482.

Walhin, J. F. (2001). Bivariate zip models. *Biometrical Journal*, 43(2):147–160.

Warrick, A. W. (1974). Time-dependent linearized infiltration. i. point sources. *Soil Science Society of America Journal*, 38(3):383–386.

Watanabe, S. and Opper, M. (2010). Asymptotic equivalence of bayes cross validation and widely applicable information criterion in singular learning theory. *Journal of Machine Learning Research*, 11(12).

Wheatcroft, E. (2021). Evaluating probabilistic forecasts of football matches: the case against the ranked probability score. *Journal of Quantitative Analysis in Sports*, 17(4):273–287.

Wilks, S. S. (1938). The large-sample distribution of the likelihood ratio for testing composite hypotheses. *Annals of Mathematical Statistics*, 9:60–62.

Winkler, R. L. (1969). Scoring rules and the evaluation of probability assessors. *Journal of the American Statistical Association*, 64(327):1073–1078.

Yang, R. and Berger, J. (1996). *A Catalog of Noninformative Priors*. Technical Report. Institute of Statistics and Decision Sciences, Duke University, Durham, NC.

Zellner, A. (1986). On assessing prior distributions and Bayesian regression analysis with g-prior distributions. *Bayesian Inference and Decision Techniques*.

Zhang, P. (1997). Comment on 'an asymptotically theory for linear model selection'. *Statistica Sinica*, 7:254–258.

Zou, H., Zeng, D., Xiao, L., and Luo, S. (2023). Bayesian inference and dynamic prediction for multivariate longitudinal and survival data. *The Annals of Applied Statistics*, 17(3):2574–2595.

Index

FSC
www.fsc.org
MIX
Papier | Fördert gute Waldnutzung
FSC® C083411

Kolibri 360 GmbH
Ferdinand-Jühlke-Straße 7
99095 Erfurt, Deutschland
produktsicherheit@kolibri360.de